To
LORI OLSON
Thanks for helping me get started in Boer goats!

and

UP LUDO
(December 18, 2005–February 20, 2006)
The sweetest little goat that ever was

Contents

1 Why Meat Goats? 1
The Pros and Cons of Raising Meat Goats • Finding and Working
with a Goat-Savvy Veterinarian • There's More Than One Way to
Raise a Goat • Which Niche?

2 Before You Begin 18
Adding Your Farm to the Equation • Availability Matters •
Do the Homework

3 Which Breed? 25
Parts of a Goat • Boers • Savannas • Kikos • Myotonics •
Tennessee Meat Goats • Composite Breeds • Spanish Goats •
Dairy Goats • Angoras • Registered, Fullblood, Purebred,
Percentage, or Crossbred?

4 Where (and Where Not) to Buy Goats 45
Individual Producers and Breeders • Which Seller? •
Production and Dispersal Sales • Sale Barns

5 Selecting Breeding Stock 54
Don't Buy Trouble • Breeding Better Boers • Look 'em Over
Before You Buy • Closing the Sale

6 Think Like a Goat 75
Range and Feeding Behavior • Social Order • Breeding Behavior:
Bucks • Breeding and Kidding Behavior: Does • Behaviors of
Newborn Kids • Goat Handling 101

SECOND EDITION

Storey's Guide to

RAISING MEAT GOATS

Managing ▪ Breeding ▪ Marketing

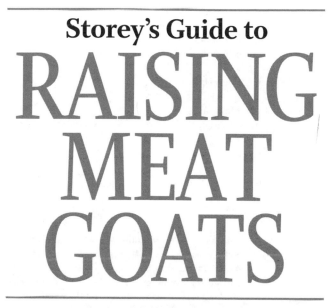

MAGGIE SAYER

Storey Publishing

*The mission of Storey Publishing is to serve our customers by
publishing practical information that encourages
personal independence in harmony with the environment.*

Edited by Lisa H. Hiley and Deborah Burns
Art direction and book design by Cynthia N. McFarland
Cover design by Kent Lew
Text production by Liseann Karandisecky

Cover photograph by © Adam Mastoon
Interior photography by © Claudia Marcus-Gurn, page 29, Sue Ann Weaver, pages 6, 215, 272, 273, and 275
Illustrations by © Elayne Sears

Indexed by Nancy D. Wood

Storey Publishing
210 MASS MoCA Way
North Adams, MA 01247
www.storey.com

Printed in the United States by Versa Press
10 9 8 7 6 5 4 3 2 1

Library of Congress Cataloging-in-Publication Data

Sayer, Maggie.
  Storey's guide to raising meat goats / by Maggie Sayer. — 2nd ed.
     p. cm.
  Includes index.
  ISBN 978-1-60342-582-7 (pbk. : alk. paper)
  ISBN 978-1-60342-583-4 (hardcover : alk. paper)
  1. Goats—United States. 2. Goat meat—United States.
  I. Title. II. Title: Guide to raising meat goats. III. Title: Raising meat goats.
SF383.4.S24 2010
636.3'913—dc22
                                                    2010029324

7 Goats on the Go 95
Stress and the Traveling Goat • Plan Ahead • Weather Kills •
Choosing a Hauling Conveyance

8 Meat Goat Housing and Facilities 103
Shelter • Pens • Feeders • Watering Devices •
Handling Facilities • Fencing

9 Feeding Meat Goats 123
The Caprine KIN (Keep It Natural) Diet and Your Goats •
The Goat's Digestive System • Goats and Poisonous Plants •
A Forage-Based Diet

10 Keeping Goats Healthy
(and What to Do When They Aren't) 136
To Be or Not to Be Your Own Vet? • Learning the Ropes •
Checking Vital Signs • Caprine Maladies You Should
Recognize • Stop Disease in Its Tracks — Vaccinate! •
Antibiotics: Good or Bad? • Epinephrine — Don't Leave
Home without It • Drenching a Goat • Medical Alternatives
for Goats

11 Parasites 101 169
Dewormer-Resistant Worms • Meet the Enemy • Other
Gastrointestinal Nematodes • Reducing Drug Resistance • Those
Other Worms • External Parasites

12 Livestock Guardians 180
Statistics Tell the Tale • Protecting the Herd

13 Breeding Meat Goats 198
Bucks Are Half the Herd • Those Does • You've Got to Be Kidding

14 Marketing Meat Goats 226
Market Goats • Breeding Stock • Show Goats •
An Unexplored Market: Working Goats

15 Promoting Your Goat Business 243
Choosing a Business Name • Web Sites Sell Goats • Industry
Advertising for Pennies • Make Business Cards Your Ace in the
Hole • Create Your Own Flyers and Brochures • 13 Additional
Easy, Low-Cost Promotion Strategies

Appendix A: DEFRA's Code of Recommendation
    for the Welfare of Goats 261

Appendix B: Photograph Your Goats 270

Appendix C: Identify Your Goats 276

Appendix D: Trim Your Goats' Hooves 282

Appendix E: A Milk Goat for the Kids 287

Appendix F: So You Want to Show Meat Goats? 295

Appendix G: Clipping for Shows 298

Appendix H: Emergency Euthanasia 299

Resources 300

Glossary 313

Index 328

1

Why Meat Goats?

GOAT PRODUCERS ALL ACROSS THE UNITED STATES are scrambling to provide America's rapidly expanding ethnic population with the goat meat they are culturally used to eating. Thousands of new producers are desperately needed to supply demand. Goat meat production is the fastest-growing segment in American agriculture. If you're seeking an easy-to-enter, small-farm agricultural enterprise with unlimited growth potential, raising meat goats is the answer to your prayers.

Consider this: Roughly 70 percent of the meat consumed globally is goat meat. Goat meat is a major staple in Hispanic, Caribbean, Mediterranean, Eastern European, Middle Eastern, African, and Southeast Asian cuisine. U.S. census figures indicate immigration to the United States increased by 20 percent between 1995 and 2000. The 2000 census tallied nearly 36 million Hispanics, 10 million Asians, and 4 million Caribbean and African immigrants residing in the United States. Add to those figures an estimated 7 million illegal aliens, more than half of whom are of Latin American descent. By 2025, Hispanics will comprise 18 percent of America's population; by 2050, one out of every four Americans will be Latino. Hispanic-Americans are major consumers of goat meat, particularly *cabrito*, the tender, pale flesh of milk-fed kids generally 4 to 10 weeks old and 20 to 25 pounds live weight.

America's Muslim population also consumes a great deal of *halal* goat meat — that is, goat meat that is slaughtered and processed according to specific guidelines. Goat is traditionally served at family celebrations such as weddings and birthdays and on religious holidays such as Eid al-Adha. When available, goat is a dietary staple for many Muslims. According to *The American*

Religious Identity Survey conducted in 2001 by the City University of New York Graduate Center, between 1991 and 2000, America's Muslim population more than doubled; by 2010, Islam is expected to become the second largest organized religion in the United States. No wonder demand for goat meat is on the rise!

Finally, health-conscious Americans who don't observe the practices of any particular ethnic group are also turning to goat meat, which boasts one-eighth the total fat found in pork and one-fifth the total fat found in beef and lamb. In fact, it's even lower in fat than skinless chicken. At the same time, goat meat is higher in protein than pork, lamb, and chicken, and has the same amount of protein as beef.

WHERE DO GOATS COME FROM?

The first domestic meat goats belonged to inhabitants of Ganj Darah, a prehistoric settlement in the Zagros Mountains of southern Kurdistan. For at least 40,000 years, men of that region had hunted the wily, wild bezoar goat before bringing kids into the fold and domesticating them about 10,000 years ago, according to Melinda A. Zeder, curator of Old World archaeology and zooarchaeology at the Smithsonian Institution's National Museum of Natural History, and Brian Hesse, a zooarchaeologist at the University of Alabama, who base their dating on goat toe bones uncovered at Ganj Darah.

In addition to hypotheses related to the Ganj Darah inhabitants, there are hundreds of references to goats in the Bible, including culinary instructions. The Hebrew Scriptures caution Hebrews, "Do not cook a kid in its mother's milk."

Meat goats also accompanied Spanish settlers and conquistadors when they came to the New World. Some acted as a traveling food supply as they followed in the wake of their owners, but also the crews of Spanish galleons "salted" islands with meat goat stock, knowing the goats would multiply and serve as a ready source of fresh meat on the men's subsequent visits. All of these goats are distant ancestors of today's Spanish goat.

To supply present demand, America imports a tremendous and rapidly increasing amount of frozen goat meat from Australia and New Zealand. According to U.S. Department of Agriculture (USDA) figures, in 1970 we imported 1.29 metric tons of goat meat. In 1990 the figure jumped to 1,361 metric tons, and in 2006 it reached a whopping 11,000 metric tons, valued at $41.8 million!

American farmers are producing more goat meat too. Slaughter of goats at USDA federally inspected slaughterhouses increased 270 percent between 1994 and 2003, when about 634,500 animals (20 million pounds) were processed. And these figures don't include on-farm slaughter or goats slaughtered at state-inspected facilities, estimated at an additional million goats per year.

When the U.S. Department of Agriculture's National Agricultural Statistic Service (NASS) conducted its first goat survey in 2004, the total number of goats in the United States was about 2.7 million, with meat goats numbering some 2.1 million. The numbers have risen steadily since then, with a slight decrease being seen from 2009 to 2010. As of January 1, 2010, there were just over 3 million goats in the United States. Of those, more than 2.5 million were meat goats.

Nevertheless, as American Meat Goat Association president Marvin Shurley said in 2004, "There are only about 2 million goats raised in the United States for meat, but the domestic market could support a herd of 15 million animals." That's a lot of goats, so if you want to raise meat goats, there's plenty of room for you in the industry.

"Because of the familiarity with goats that many cultures have, it is no surprise that as those cultures become established in the United States, they demand those foods with which they are familiar. Therefore, considering the popularity of goats and goat's milk, the growth in the goat industry will continue as long as the ethnic population within the United States continues to grow."

— The Goat Industry: Structure, Concentration, Demand and Growth
(U.S. Department of Agriculture, 2005)

DON'T CALL IT GOAT MEAT!

In the United States, most goat meat is marketed as *chevon,* a term coined by combining the French words *chèvre* (goat) and *mouton* (sheep) and used to describe the meat of older kids and adult goats. The pale, tender flesh of milk-fed kids is called *cabrito* (Spanish for "little goat") or occasionally *capretto* (Italian for "little goat" and a term more commonly used in Europe, Great Britain, and Australia). Some buyers simply call it *goat.*

In the late 1980s, however, the Florida Agricultural Market Research Center, associated with the University of Florida at Gainesville, conducted studies to evaluate consumers' perceptions of goat meat, and 80 percent of interviewees suggested changing the name to something that would connote a meat product rather than an animal (in the manner of *veal* instead of baby cow, *pork* instead of pig meat, and *beef* instead of cow meat). When offered their choice of names, 40.3 percent favored chevon, 41.4 percent preferred cabrito, and 18.3 percent were satisfied with goat meat.

The conclusion of this research? Says Dr. Robert L. Degner, in *Should You Market Chevon, Cabrito or Goat Meat?* (www. agmarketing.ifas.ufl.edu/pubs/1990s/GOAT.pdf): "If you are satisfied to limit your market to people familiar with goat, 'goat meat' is fine. However, if you want to build a larger market by appealing to the masses who have never tried this delicacy, your chances of success will be increased by calling it chevon, cabrito, or other exotic-sounding name."

"National estimates based on import data indicate that the United States' supply of goats is deficient — more than 500,000 additional goats are required to meet the country's current demand for goat meat."

— Sandra G. Soliaman, Ph.D., associate professor of animal and poultry sciences at Tuskegee University, in *Outlook for a Small Farm Meat Goat Industry for California* (University of California Small Farm Center, 2006)

The Pros and Cons of Raising Meat Goats

There are plenty of great reasons to raise meat goats:

- Goat meat is in high demand and prices are on the rise. It will be decades at least before supply catches up with demand. You may have to advertise and promote to establish yourself locally, but due to America's ethnic population explosion, the overall national market is established and rapidly growing.
- It doesn't cost a fortune to get started in goat farming. Compared to other livestock species, goats are inexpensive to buy and feed. They don't require fancy housing; simple, three-sided loafing sheds are sufficient in many locales and most existing farm structures can be easily refitted for goats.
- Unlike cattle, goats offer a quick return on initial investment. Their gestation period is five months, compared with nine months for cattle. First-time moms usually give birth to a single kid, but after that, twins are the norm, triplets and quads are common, and quints are not unheard of. Market (slaughter) kids are sold at weaning time, three to four short months after birth. And since most meat breeds come into heat year-round, it's easily possible to achieve three crops of kids in two years.
- Because goats are small and relatively easy to handle, they're ideal for farmers intimidated by or physically unable to cope with larger livestock. Their size makes expensive handling equipment unnecessary in small-farm settings and optional where larger numbers of goats must be vaccinated, dewormed, or sorted.
- Goats are less labor-intensive than most other types of livestock, making part-time goat farming a doable venture.
- Goats require less pasture (and supplementary feeding) than cattle, making goat farming a best-bet scenario for hobby farmers and for sustainable small-farm operations. Six to eight goats require the same amount of improved pasture as a single horse or cow; on rough, brushy graze the ratio is closer to nine to eleven goats to one cow. Furthermore, since goats prefer browse (brush, saplings, briars, and weeds) over grass, they can be pastured with cattle or horses (one or two goats per cow or horse) without any impact on the larger animals' grazing. At the same time, goats effectively control brush and weeds that would otherwise compete with pasture grasses grazing species prefer.
- Finally, goats are intelligent, inquisitive creatures that are just plain fun to have around. A pasture full of playful kids has entertainment value beyond anything you'll see on TV.

Nevertheless, raising meat goats isn't a stroll in the park. Consider these points before committing to goat farming:

- To keep them contained and to keep out predators, goats require elaborate fencing. In fact, fencing is the major start-up cost for most new producers. Cattle-proof barbed-wire fencing can be upgraded for goats by adding strands of barbed or electric wire, but tall, heavy-duty cattle panels and woven wire designed specifically for goats are far better choices. Because goats are escape artists, even cattle panels and goat fencing need additional strands of electric or barbed wire to make them goat-tight.

- Goats are vulnerable to predators such as coyotes; free-roaming dogs; wolves; bears; and large bobcats, cougars, and other sizable felines. Predator-resistant fencing is a goat farmer's first line of defense against these intruders. Night-penning goats near human habitation helps too. Established producers strongly advocate keeping livestock guardian dogs (LGDs) with goats; llamas or donkeys can be used in lieu of dogs in certain situations. (See chapter 12 for more on these guardians.) Don't assume horned goats can protect themselves; they can't. Providing adequate protection from predators is a major part of farming goats.

- Goats are extremely susceptible to internal parasite infestation, yet as worms become resistant to certain chemicals used in goat anthelmintics, deworming medications are becoming increasingly ineffective. Goats must be dewormed on an ongoing basis and must be tested to make certain the deworming medication used is still working.

Two old goats: MAC Goats Chief Forty-five ("Chiefee") offers some advice.

- Goat-savvy veterinarians are few and far between, so most producers must learn to vaccinate, diagnose, and doctor their goats themselves. This is sometimes not for the faint of heart; if you aren't willing to do it, find a goat vet before you commit to raising these creatures.
- For the first few years in the goat business, goat farmers can count on making mistakes (sometimes serious ones), even when they have help from vets and goat-wise mentors. Once you have a handle on feeding, kidding, and parasite control, goat keeping is relatively easy, but until then, you'll lose plenty of sleep and probably some goats. There's no way around it; it's part of the learning curve.

WHERE THE GOATS ARE

According to the most recent National Agricultural Statistics Service figures, America's top-20 goat-producing states in 2009 and 2010 were:

National Total	2,549,000	2,538,000
By State	2009	2010
Texas	980,000	990,000
Tennessee	134,000	125,000
North Carolina	86,000	95,000
California	95,000	93,000
Oklahoma	105,000	90,000
Missouri	86,500	84,600
Kentucky	79,000	79,000
Georgia	80,500	79,000
Florida	55,000	60,000
Alabama	65,000	60,000
Arkansas	47,000	53,000
Virginia	62,000	52,000
Ohio	55,500	50,000
Pennsylvania	42,000	42,000
Kansas	50,000	42,000
South Carolina	38,000	39,100
Colorado	34,000	38,000
New York	27,000	35,000
Indiana	32,000	33,500
Oregon	28,000	30,000

"Prices for slaughter kids have risen from $.75 per pound in 1996 to approximately $1.36 per pound in 2005, resulting in an 81 percent increase. From 1996 through 2003 domestic production of goat meat increased 81 percent, imported goat meat by 139 percent, and total meat consumption by 97 percent."

— Marvin Shurley, president of the American Meat Goat Association, in "The U.S. Meat Goat Industry: Past, Present, Future," presented at the Gathering of Goat Producers IV

Finding and Working with a Goat-Savvy Veterinarian

Unless they're goat owners themselves, relatively few vets have hands-on experience with these animals. If you're lucky enough to find one who does, treat him or her like spun gold. If you can't find a vet with goat experience, try to find one who knows sheep (sheep medicine and goat medicine are very similar) or a vet who is willing to work with you and research your goats' needs as problems arise.

One way to find a veterinarian with goat experience is through the American Association of Small Ruminant Practitioners website (see Resources). It posts a state-by-state list of such veterinarians. Perhaps the best way to find a vet, however, is to ask other goat producers in your locale for their recommendations; ask dairy and fiber goat owners too. Also ask for names of vets they wouldn't use and why. Narrow it down to two to four vets most goat owners like, then call these doctors' offices and ask these pertinent questions:

- Does the doctor make routine farm visits or must you take your goats to the vet clinic for treatment?
- What facilities are available if patients must be left overnight (or longer)?
- Will the vet come to your farm in an emergency? What about after-hours, weekend, or holiday emergencies? To whom does the vet refer clients when he or she is unavailable?
- How many veterinarians are associated with the practice? If more than one, can clients stipulate which vet they want to attend to their animals?

- If you phone the clinic with a problem, will the receptionist or answering service help connect you with the vet him- or herself (within reason) or will desk personnel relay your concerns and return your call?
- How is payment handled? Is up-front payment required for every call or is the vet willing to bill clients? Are credit cards acceptable? Does the clinic offer payment plans?

If you like what you hear, arrange for a time to visit the clinic. During your visit, note whether the staff is friendly and knowledgeable and whether they'll allow you to speak with the vet. Remember, desk personnel and vet technicians are your link to the vet, so if they make you feel uncomfortable, keep looking. In your tour of the facilities, note whether large-animal lodgings are safe, clean, and arranged so patients can't physically interact with one another. Note what sort of feed patients are eating and ask if you can provide your own feed if you prefer. Also note whether clean water is readily available.

If bedside manner matters to you, in your meeting with the vet, determine how comfortable you are with his or her demeanor. Because you'll likely be performing some routine care on your goats, determine how the vet feels about clients taking on health care procedures such as treating minor ailments and giving their animals shots. Ask the vet if he or she will dispense prescription drugs such as Banamine and Nuflor once it's been determined that you're competent to administer them. Finally, determine how the vet feels about clients researching problems in books or online and bringing the results to him or her for perusal.

Remember that whatever you ask, you should never be made to feel inadequate or stupid. They're your goats, after all, and you'll be footing the bill; if you're made to feel inferior, take your business elsewhere.

Once you've selected a vet, don't wait for an emergency to use his or her services. Schedule a routine farm visit. Does the vet arrive promptly, or, if he or she is held up, does someone from the vet's office phone to inform you of delays? How does the vet interact with your goats? Are you comfortable with his or her attitude and medical care? If so, congratulations—you've found your goats a vet!

In turn, it's only fair to treat your new vet right:

- Have your goats confined and ready for treatment when the vet arrives. Chasing wayward goats across a 50-acre field is not in his or her job description.

- Learn to handle emergencies on your own until the vet arrives. A country vet may be 50 miles away on another farm visit when you most need him or her. Make certain your cell or cordless phone works in the barn in case the vet needs to talk you through a procedure. Stock well-equipped first-aid and birthing kits and know how to use them.
- Don't, however, wait until a minor problem escalates into an after-hours or weekend emergency. Know what you can and cannot do yourself and involve the vet as soon as you're clearly out of your depth.
- Be there during the vet's visit. Your animals know you and they'll behave better if you're there to help. If you don't understand a treatment, ask questions. If a follow-up treatment you'll have to administer entails detailed instructions, write them down and follow them to the letter.
- Furnish a comfortable, weatherproof, well-lit place for your vet to work. Provide any restraints necessary to secure your goats.
- A cold beverage on a sweltering summer afternoon or a steaming cup of coffee in the winter is always appreciated. And always settle your bill when payment is due. Good goat vets are hard to find.

There's More Than One Way to Raise a Goat

Goats are amazingly adaptable creatures and can be raised under widely varying conditions. Large-scale goat ranchers often raise goats on vast tracts of western rangeland or southern brush. Goats are expected to fend for themselves with little or no supplementary feeding and with minimal human intervention. Livestock guardian dogs protect these goats from predation, but no one trims their hooves, doctors sick goats, or assists at kidding time. Goats are wild, so they're run through handling systems for occasional deworming and annual vaccinations. Most kids are collected at weaning time and shipped to market; better doelings are sometimes retained as herd replacements or raised to be sold as adult breeding stock. Nowadays, does in these herds are mainly hardy crossbreds produced by breeding Spanish does to Boer, Kiko, or improved Myotonic bucks. Herds of 1,000 to 3,000 does are the norm.

Other producers utilize smaller pastures and woodlands for smaller herds. Some of them cross-fence pastures into paddocks as small as ¼ acre apiece and practice rotational grazing by moving their goats to fresh graze or browse every few days. Most (but not all) smaller-scale goat farmers routinely trim hooves, deworm on schedule, and monitor their does when they give birth. Since these goats are handled more often than those in the first group, they're generally tamer than their range-raised cousins.

"Since 1994, there has been an increase in goat meat imports at a rate of 30 percent annually and domestic slaughter has more than doubled since 1980. . . . Of all red-meat species, only goat meat consumption has increased significantly over the last two decades."

— Doolarie Singh-Knights and Marlon Knights, in *Feasibility of Goat Production in West Virginia* (West Virginia University Bulletin 728, 2005)

A small percentage of meat goat producers dry-lot their goats. Instead of turning them out to graze or browse, dry-lot goats are fed hay and grain year-round. Because purchased feed is expensive and goats tend to waste a lot of hay, this management system isn't feasible for breeders of market kids. Extra labor is required for feeding, manure cleanup, and other daily chores, so it isn't the ideal way to raise high-ticket goats either. Goats raised in dry lots never learn to go out and graze for food; boredom and crowding increase aggression and in-fighting; and goats in dry lots tend to get fat. Fat, under-exercised does are prone to pregnancy toxemia and a host of other kidding maladies. This system is doable, but it isn't very profitable and it's not in your goats' best interests.

Which Niche?

While raising market kids for slaughter is the main thrust of the meat goat industry, there are several other possible enterprises from which to choose. For now, let's take a brief look at each; we'll discuss them at length later on.

Raising Market Goats

Goat producers in this niche raise kids targeted to sell prior to ethnic holidays; goats to sell at auction or to meat brokers year-round; or goats to direct-market straight from the farm. Profits vary widely, depending on the cost of raising kids to market size and on current market prices.

Kids for Preholiday Sales

Because preholiday prices are often the highest of the year, raising kids for ethnic holiday feasts is arguably the most profitable niche market. To reap that profit, however, a producer must know when each holiday occurs in order to breed does and wean kids at precisely the right times. He or she must also understand the needs of ethnic consumers and supply the type of goats potential buyers prefer. For instance, buyers for Eastern Orthodox Easter celebrations want plump (not fat), milk-fed kids that tip the scale at approximately 35 pounds live weight; Hispanic families planning to barbecue a Cinco de Mayo kid want a milk-fed youngster weighing 20 to 35 pounds live weight. Goat stew for Jamaican Independence Day calls for young, smelly bucks, 60 to 80 pounds live weight. And Muslim customers buying for an Eid al-Adha feast require blemish-free, uncastrated yearlings with one set of permanent teeth.

WHO WANTS WHAT TYPE OF GOAT AND WHEN?

Holiday	Religion	Type of Goat Desired	2011
Beginning of Eid al-Adha (holiday lasts four days; exact date depends on sighting of moon's phase)	Muslim	Blemish-free yearlings (uncastrated; no broken horns, no scars, no wounds)	October 18
Start of Ramadan	Muslim	Male (castrated or not) or female kids 12 months old or younger; optimum live weight of 60 lb.	August 1
Eid al-Fitr	Muslim	Male (castrated or not) or female kids 12 months old or younger; optimum live weight of 60 lb.	August 11
Western Easter	Christian	Plump (not fat), milk-fed kids three months of age or younger; live weight 20–50 pounds (ideal is 30 pounds)	April 24
Orthodox Easter	Christian	Same as for Western Easter (ideal is 35 pounds)	April 24
Passover	Jewish	Fat, milk-fed kids; live weight 30–55 pounds	March 19–26
Cinco de Mayo (Mexico's Independence Day)	Interfaith (Hispanic/ secular)	Milk-fed kids, live weight 20–35 pounds	May 5
Fourth of July	Interfaith (U.S./ secular)	Cabrito-sized kids 4–10 weeks old; live weight 20–25 pounds	July 4

"Meat goat production is one of the oldest and most widespread of the animal enterprises in the world. At this point, it is the fastest growing animal enterprise in the United States."

— *Marketing of Meat Goats* (University of Arkansas
Division of Agriculture, 2005)

2012	2013	2014	2015
October 26	October 5	September 24	September 13
July 20	July 9	June 29	June 18
August 19	August 8	July 29	July 18
April 8	March 31	April 20	April 5
April 15	May 5	April 20	April 12
March 7–14	March 26–April 2	March 15–22	March 4–11
May 5	May 5	May 5	May 5
July 4	July 4	July 4	July 4

Goats for the Open Market

Some producers prefer to market slaughter goats year-round. This works well for goat ranchers living near ethnic communities where goat eating is the norm. These producers usually sell goats through buying stations and auction and to meat brokers.

On-Farm Goat Sales

Many goat producers living near metro areas and college towns with large ethnic populations sell meat goats straight from the farm. In some cases, customers visit the producer, select a goat, and pay for it, then slaughter the goat on the producer's farm. In other cases, customers purchase a goat—in person or by phone—and have it delivered to a slaughterhouse for processing. Most

ETHNIC GROUPS' NONHOLIDAY GOAT MEAT PREFERENCES*

Ethnic Group	Goat Meat Preference
Caribbean Islanders	Mature goats, especially bucks; goat heads; preparation: head removed, feet on, singed
Chinese and Koreans	Live weight 60–80 pounds; preparation: head on, feet removed, hide on but hairless (scalded and scraped)
Greeks	Young, tender kids; live weight 30–40 pounds; 8–16 weeks old
Hispanics	Varied preferences: milk-fed kids, live weight 15–25 pounds; 4–10 weeks old (cabrito); preparation: head on, feet removed, skinless
Italians	Milk-fed kids; live weight 18–28 pounds; 4–12 weeks old (capretto)
Jews	High-quality kids; live weight 20–40 pounds; preparation: kosher slaughtering
Muslims	Lean kids; live weight 50–70 pounds; under 12–14 months old; preparation: halal slaughtering
Somalis	Live weight 40–60 pounds; preparation: goat livers, kidneys; usually require halal slaughtering
West Africans	Varied preferences; will often prepare mature goats, especially bucks; preparation: head removed, feet on, singed

*Adapted from Andrew Larson and Evelyn Thompson, *Niche and Ethnic Markets for Goat Meat in Illinois* (Urbana: University of Illinois Extension); Evelyn Thompson, Gina Backes, and Dar Knipe, *Marketmaker Chicago Ethnic Markets: Goat* (Urbana: University of Illinois Extension); and Doolarie Singh-Knights and Marlon Knights, *Feasibility of Goat Production in West Virginia* (Morgantown: West Virginia University).

producers who sell this way factor slaughterhouse delivery into their prices. In either scenario, by eliminating the middlemen, sellers create extra value for their customers and earn a tidy profit for themselves.

Breeding Stock

Producers who choose not to raise goats for a terminal market find their niche in raising bucks and does to sell as breeding stock. Many goat ranchers make a nice living breeding seed stock for entry-level entrepreneurs and replacement does for market kid producers. Ranchers in this category also market excess buck kids as meat.

While such enterprises are frequently large-scale, no- or low-maintenance range operations that are ready and eager to sell in lots of several hundred or several thousand does, this facet of the industry can also work for small-scale goat raisers, particularly in areas where the meat goat industry is just gaining a toehold and healthy young does are in high demand.

These operations raise "commercial goats": hardy, crossbred kid-making machines produced by breeding fullblood or high-percentage Boer, Kiko, or improved Myotonic bucks to Spanish, percentage Boer or Kiko, and meat and dairy breed mixed does. Healthy commercial does of these mixes fetch from $80 to $300 each, based on genetics and availability.

Fullblood Bucks for Commercial Operations

Most commercial goat raisers buy high-quality fullblood meat goat bucks to head their herds. Because most allot one buck for every 20 or 30 does, breeding these bucks can be a profitable venture; so can selling fullblood does and doelings to other breeders who are eager to breed their stock.

Breeders in this category select for commercial qualities such as ability to gain weight rapidly, meaty conformation, good udders, and mothering ability rather than show-ring flair. Some choose goats of popular bloodlines, while others don't care much about lineage. Prices for commercial bucks vary greatly depending on breed, quality, pedigree, and availability. Performance-tested Boer and Kiko bucks tend to sell at higher prices. Fullblood commercial-quality Boer bucks can fetch as little as $300 and as much as or more than $1,500. Kikos as well as less-common breeds such as Savannas and improved Myotonics (Tennessee Meat Goats) sell at the higher end of this range. Culls are generally sold without registration papers or sent to market as meat.

CHOOSING NOT TO SELL GOATS FOR MEAT

If you want to raise meat goats but not market your kids for meat, you can turn a profit selling high-quality, fullblood breeding stock or show goats. To avoid producing large numbers of culls, learn all you can about your breed before you buy your first goats. Start with proven, top-of-the-line goats of popular bloodlines and breed your best to your best. Understand going in, however, that no matter how outstanding your goats may be, to maintain a high-quality herd, some culling is inevitable, so plan ahead for the future of your culls.

Goats culled for lack of show qualities but that are otherwise decent examples of their breed can be marketed as breeding stock for commercial goat producers. Castrate lesser-quality bucklings; don't allow them to breed more culls for somebody else. Raise less-than-stellar youngsters to adulthood, train them as harness or pack goats, and sell them to working-goat enthusiasts. Donate cull kids and old does as pets to country nursing homes, group homes, and residential youth programs. Your less-than-perfect goats can make a difference in people's lives. Sell or donate doelings, sans registration papers, to 4-H youngsters to raise as no-kill meat goat doeling projects. Place culled goats, without registration papers, in approved pet homes. Humane society personnel and rescue group coordinators are usually happy to help livestock owners set up their own placement programs.

Fullblood Show Stock

Boers are the kings and queens of the meat goat show world. The three organizations that register Boer goats sanction hundreds of shows featuring thousands of classes for fullblood and percentage Boer goats each and every year. Competition is stiff; it takes a top-flight goat to bring home the blues. Producers willing to show, cull ruthlessly, advertise, and promote, promote, promote can earn a tidy income breeding the crème de la crème of the Boer goat world. Show-quality bucks bring $1,500 and up (way, way up) and show does sometimes sell for even more.

A small herd of meat goats can be produced on 10 to 15 acres of pasture-land and can fit into more than 90% of the U.S.'s farmsteads and enhance small farm diversity and profitability.

— Sandra G. Solaiman, PhD, PAS, in "Assessment of the Meat Goat
Industry and Future Outlook for U.S. Small Farms"
(Tuskegee University, August 2007)

4-H and FFA Wethers

In many states, 4-H meat goat projects are skyrocketing in popularity. Some-one has to provide these youngsters with goats. Savvy commercial goat ranchers castrate their best buck kids to sell as 4-H and FFA market wether projects. (A *wether* is a castrated male goat.) Fullblood breeders sometimes sell better-quality culls as project goats too. Some market project kids by the pound, others by the head, depending on how it's usually done in their locale.

The going price for newly weaned youth project wethers range from $100 to $750 or more, depending on quality, locale, and availability.

The goat business is a wide-open market; thousands more goat producers are needed — and they're needed right now. There's a place in the goat industry marked with your name. If you're ready to claim that place, read on!

YOUTH GOAT PROJECTS ARE GROWING

"Market steers increased by 7.78 percent during [1998–2000], market swine increased by 2.67 percent, market lambs decreased by 6.63 percent, and meat goats increased by 55.15 percent. That kind of tells you right there that it is the fastest growing of the livestock projects or youth programs by far."

— Warren Thingpen, Bandera County (Texas) Extension agent,
commenting on a statewide survey of youth livestock projects
in the October 30, 2001, issue of *AgNews*

2

Before You Begin

GOATS AREN'T THE HARDY, TIN CAN–EATING CRITTERS of movie and cartoon fame. To succeed in a goat business, even a small one, you'll need goat-resistant fences, a well-stocked medicine chest, and a good bit of know-how — not to mention healthy foundation animals.

Before you proceed, be sure to investigate the main segments of the meat goat industry, do your homework, and be reasonably certain which ones your business is going to serve. The does you buy to produce commercial slaughter kids may cost $150 each, while a show-winning, fullblood Boer doe can easily set you back $5,000. You'll save considerable amounts of time, frustration, and money if you know where you're going *before* you buy foundation stock.

Unless you have prior experience with goats (and, to a lesser extent, other ruminant livestock such as sheep and cattle), you'd be wise to ease into production rather than jumping in feetfirst. While goats aren't difficult to keep, there is a learning curve new goat keepers must master. You'll make mistakes — they're inevitable — so it's better (and less costly, financially and emotionally) to learn the nuances of kidding; predator, parasite, and disease control; and the fine art of nursing sick goats with a group of 10 or 20 goats than it is with a commercial herd of 1,000 does.

Once you've mastered the art of goat keeping, you'll be ready to expand your herd. How many goats you'll need to show a profit depends on which facet of the goat industry you plan to serve (a fullblood show stock breeder might own 20 to 100 head while a commercial producer of slaughter kids runs 2,000 does), your market, climate, facilities, available help, and more.

Adding Your Farm to the Equation

When deciding what type of goat enterprise to engage in and how many goats to buy, take into consideration such factors as how much time you plan to devote to your operation, how much help you can muster when you need it, and your farm's physical amenities such as climate, shelter, fencing, and suitability of pasture or browse.

Labor

Unless you take a minimum-intervention approach to herd management (and some very successful commercial goat producers do), you'll need to allow time (and likely recruit help) to manage your herd at kidding time or when health problems occur. If someone in your family is ready to assist, so much the better. A nearby goat-keeping mentor may be enough if your herd is small. Otherwise, it's especially important to ease into goats rather that leaping in feetfirst with a huge herd you can't manage by yourself. Trimming 1,200 hooves or treating 300 does for pinkeye isn't something you want to tackle on your own.

Climate

Though climate-related difficulties can be circumvented, it's easiest to pick goats adapted to the climate where you live. For instance, while Boers are the goat of choice in most meat herds, ultra-hardy, adaptable Kikos or Boer-Kiko crosses might do better in colder northern climes or in the sizzling and muggy deep South. Whichever breed you choose, don't transplant goats from vastly different climates and expect them to immediately thrive in your own. When possible, bring them in during your mildest season (northern goats to Texas during the winter months, Texas goats to the North in the summer) to give them ample time to adapt.

Facilities

Goats don't require fancy housing (we'll talk about this again in chapter 8), but they do need shelter from direct blazing sun, snow, and rain. Meat goats fare best in basic, loafing-type structures but most any existing structures can be modified to house goats.

We'll discuss fencing in chapter 8, but for now, suffice it to say that you *don't* want to haul home goats — not even just a few — until secure goat fencing is in place. Goats live for the thrill of escaping their fences, and predators savor tasty goat. If you have nearby neighbors, your escapees will raid their gardens and tap dance on the hoods of their cars. If you think we're exaggerating, speak to goat keepers at random; they'll back us up on this, I assure you.

You can dry-lot goats if you absolutely must, but it's neither a profitable nor a healthy long-term way to keep them. Meat goats have taken America's hill country by storm because goats prefer browse to lush green grass. Rocky, partially wooded land that can't carry cattle is ideal for raising meat goats.

Availability Matters

Before choosing a breed, decide if you want what's locally available or something unique to your locale. This isn't terribly important when raising commercial slaughter goats, but it is for breeding and show stock producers.

If you choose what's popular, you can usually find replacement stock without traveling too far from home; however, you'll have more competition when marketing goats of your own.

If you choose a lesser-known breed, you may have to buy foundation and replacement goats from afar and incur traveling or commercial hauling expenses to get them home. The good thing is that other producers in your area who already fancy this breed will come to you to buy stock; the bad thing is that you may have to promote more than the norm to market breeding stock locally.

Budding commercial goat raisers should discuss local markets with established goat producers to find out which type of slaughter goat sells best. For instance, in many parts of the country, white kids with colored heads fetch higher prices than the rest; in others, ethnic buyers prefer a variety of colors so they can choose goats colored like the ones they had at home.

Do the Homework

It's easy to get stung when entering into any new business enterprise, so before buying anything, educate yourself and eliminate some of the risk.

Talk with as many established breeders as you can. Goat people are friendly and they love to talk about goats. Find a local mentor if you can—nothing is better in a crisis than knowing someone you can call to come and help you out.

Contact your County Extension agent to find out if there's a goat organization in your locale. If there is, join and attend its meetings. If there are meat goat associations in your state or region, be sure to join those as well.

Boot up your computer and visit the educational websites listed in Resources, buy a three-ring binder, print out the articles you like best, and compile a goat reference book tailor-made for your needs. Or download PDF files and save them to CD or portable drive so they'll be ready at hand when

NEW GOAT OWNER'S CHECKLIST

Are you really ready to take possession of your first goats? Use this handy checklist and be certain. You will need:

- ☐ A basic first-aid kit (a fully packed kit is better). Read about it in chapter 10.
- ☐ A well-stocked kidding kit. This is covered in chapter 13.
- ☐ Safe shelters (and bedding for any goats kept indoors). See chapter 8.
- ☐ Properly fenced pastures or a large holding area to use until your fences are in. We'll also discuss fencing in chapter 8.
- ☐ A quarantine area at least 50 feet from any other goats. We'll talk about this in chapter 4.
- ☐ Handling facilities (if you need them). See chapters 6 and 8.
- ☐ Predator protection: A safe, close-to-the-house area in which you can pen your goats at night; livestock guardian dogs or guard llamas or donkeys. We'll cover these bases in chapters 8 and 12.
- ☐ Quality hay, grain, minerals, and possibly protein blocks. We'll discuss feeding goats in chapter 9.
- ☐ Feeders, water containers, and a clean water source. All are covered in chapter 8.
- ☐ Hoof-trimming tools (you'll probably need them sooner rather than later). We'll learn which you need and how to use them in appendix D.
- ☐ Halters, collars, and leads (don't haul goats by their horns!).
- ☐ Phone numbers of at least three veterinarians who are willing to treat goats (not all will). If the one you want isn't available, know of at least two other vets you can call when you desperately need one.
- ☐ Phone numbers of at least three knowledgeable goat owners you can call in an emergency; barring that, investigate free services like Goat 911 (it's listed in Resources), and be sure you know how to use them.

you need them. And by all means subscribe to as many goat listservs as you can handle. YahooGroups (www.yahoogroups.com) hosts more than 1,000 goat-specific e-mail lists, and membership in these groups is free. Listservs are rich and varied sources of information you simply can't find anyplace else, so join a bunch of them, read, and ask questions. New ones are added to Yahoo's lineup every day. Do a search using your breed's name or simply the words *meat goat*; you'll be surprised how many great lists you'll find.

Producers Profile

Lee and Connie Reynolds, Autumn Farm Boers
Ravenswood, West Virginia

With her husband, Lee, Connie has been raising goats in the hill country of West Virginia for the past 16 years. Connie also writes *The Nannie Berries*, wildly funny stories about life on a West Virginia goat farm. See Resources for where to read them online (you'll be very glad you did).

Q *Connie, how and why did you and Lee get started in Boer goats?*

A Lee and I have been married over 32 years and throughout those years we've always had a goat or two wandering around. It's all Lee's fault: two months into our marriage, he brought home a bottle baby. She was a girl of mixed heritage whom we ingenuously named Nanny. Thus began our fascination with goats.

We lived in many places in the United States, but eventually came home to West Virginia. We purchased a farm and started stocking it with horses and, of course, a few goats. West Virginia has marvelous hills and we lived in a place where, in order to see the sun, we had to stand between the hills and look straight up. One day Lee rolled a tractor on one of those hills while he was trying to brush-hog. The emergency room doctor said he was lucky because he'd only fractured every bone in his body. I looked at Lee and said, "We need more goats. Let those goats do the brush hogging." And on that fateful day he agreed.

So we increased our Nubian herd, added LaManchas and French Alpines, and then bought a herd of Angoras. We even had some Pygmies, but the little stinkers kept squirting through the fence and going where they weren't wanted, so we reluctantly parted with them.

Then in 1993, Lee saw some pictures in a *Southern States* magazine of a new breed coming to America: Boers. The muscles on those Boers, the red heads — we both just fell in love. At that time the cost of Boers was unreal. People were mortgaging their farms to get a registered buck. Still, we went into partnership with a friend and bought a registered fullblood baby buck. We sold our Nubian and Angora bucks and put that Boer buck with our girls. He soon proved that anyone who says a Boer is a pampered pet that can't survive is wrong. This buck came along when people in our area thought deworming a goat more than once a year would poison it. No one treated for coccidia, no one gave shots to prevent illnesses, and a goat wasn't given grain unless it was a producing dairy doe. But this expensive baby Boer buck thrived in the brush and on what I would consider now a lack of care, although that was considered normal back then. That Boer buck never missed a beat and he produced a lot of meaty little kids. That was our start in Boer goats.

Q *Do you feel Boers are better than the other meat goat breeds?*

A In every breed you'll find individuals who are stronger or weaker than the others. But I haven't found one breed of goat to be tougher than any other breed. I just happen to prefer Boers.

Q *What sort of Boers do you produce?*

A We usually keep at least one hundred head of goats on our farm. We have Boer crosses, purebreds, and fullbloods. We sell goats for 4-H projects and breeding stock, and we sell commercial goats. We're into raising the fastest-growing, meatiest goat we can possibly raise. We try to hit a lot of different markets so if one is down one year, the others will carry the slack. The main requirement our goats have to meet is that they pay for themselves. If I can make a little extra money to add new run-in sheds, build a new barn, or buy a tractor, that's quite all right too.

Q *Do you think this is a good time for folks to break into raising meat goats?*

A The goat market is good and growing. Nowadays there are more Boers around here than you can shake a stick at, so it's easier and not as costly for a person to get into Boers, and there is a nice variety of bloodlines to select from.

Q *What advice would you give new meat goat producers?*

A Be prepared to do your own deworming, shots, and goat doctoring. Most vets still aren't that interested in goats, so learn which end that digital thermometer goes in, what dewormers you can use and how much to give, and how to give a shot. I used to bawl like a baby every time I had to give a shot, both before the shot, during, and afterward. Once I saw that both the goat and I survived, it got easier each time. Why, I'm even brave enough now to give shots to cats. They're something I consider very dangerous. Cats are like a gun: you don't want to aim one at your face — and now I can give those dangerous animals shots, if Lee holds them down.

Another piece of advice: For goodness' sake, if you have a first-time kidder, keep an eye on her. No telling what that first-time little innocent will pull when she kids. If you get into raising 4-H projects to sell around here, you've got to be willing to kid in winter and provide a barn and heat lamps and be willing to be there as each kid is born, because left alone in the temperatures in our area, the kid will freeze and die. You need to be his cheerleading department to make sure he gets up and nurses quickly, getting that good warm colostrum, and show him where the nice warm heat lamp is at.

Oddly enough, the better care you give your goats, the better you and your goats will be. Oh, I don't mean making sure they have a color TV in every stall, but simply giving them the basics such as deworming, good feed that fits a goat's requirements, goat minerals, goat protein blocks, good hay, that sort of thing. The extra cost is paid back in healthier goats that need less doctoring, in every doe giving you at least twins, in every buck being very productive. Some people sit down and figure out the number of goats they can lose and still make money. To me, that's totally unacceptable. Each lost goat is a loss in profit. It comes down to management and figuring out how to care for the herd without going into the hole doing it. Learn all you can and then you won't spend as much time doctoring animals or losing goats and saying it's all their fault. If you're going to do a job, do it right, and then you can really enjoy it.

3

Which Breed?

IF YOU PLAN TO RAISE SHOW AND BREEDING stock, you'll need to choose a breed before buying any goats. Even if you raise commercial meat goats, you'll probably want to use a fullblood, purebred, or high percentage buck for breeding, and because he'll have a huge impact on your kids' marketability, you'll want to choose the best breed for your needs. Though Boers revolutionized the meat goat industry and are still the standard in most locales, other breeds are rapidly gaining in popularity. This chapter introduces the current major players.

Parts of a Goat

Before we begin any discussion of breed, as well as later discussions of goat health and care, it's important to identify the parts of a meat goat. The same terms are used to describe body parts for dairy goats and meat goats, with the exception of the *twist*: that bulging muscle at the rear end of a meat goat that is akin to a buttock. See diagram on next page.

Boers

The Boer is the quintessential meat goat. Both sexes are long, large framed, and massively muscled. Large size; meaty build; long, pendulous ears; and a convex (Roman-nosed) facial profile are the Boer goat's trademarks, as are the rolls of wrinkled skin over a Boer buck's shoulders. Most are white with a red or brown head, although Boers come in black, red, and paint (spotted), too.

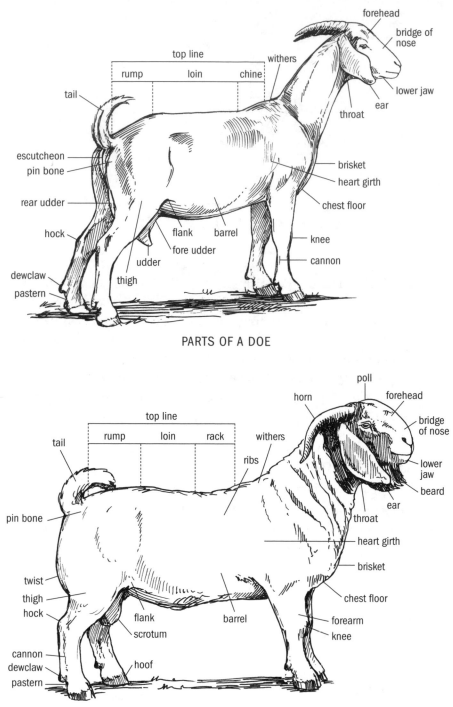

PARTS OF A DOE

PARTS OF A BUCK

BREED: Boer

Origin: South Africa

Color: Traditional (white with a light brown to dark red head), black traditional (white with a black head), paint (spotted), red, black; most Boers have white facial markings.

Description: Large size; long, broad, and muscular build; long, pendulous ears; convex facial profile; swept-back horns; short coat; bucks have rolls of loose skin on shoulders.

History

Dutch farmers settled this breed's native South African homeland beginning in the mid- to late 1600s; *boer* in Dutch means "farmer" and *boerbok*, "farmer's goat." The breed's exact origin is unclear, but most breed historians believe today's Boers descend from indigenous African goats crossed with European dairy breeds (including English Anglo-Nubians of Indian ancestry).

Modern Boer goats were developed in the Eastern Cape province of South Africa during the early 1900s. In 1959, the Boer Goat Breeders Association of South Africa was formed and in 1970 Boers were incorporated into the South African National Mutton Sheep and Goat Performance Testing Scheme, making Boers the first goats in the world to be scientifically performance-tested for meat production.

Boers weren't initially sold outside of South Africa. In the late 1980s, however, a contingent of New Zealand breeders purchased embryos in Zimbabwe, took them home, and implanted them in recipient does: thus began New Zealand's love affair with Boer goats.

In the early 1990s, frozen embryos from New Zealand-based Landcorp Corporation were shipped to Olds College in Olds, Alberta, Canada, and implanted in recipient does. The following year, embryos were shipped directly from South Africa to Canadian recipient herds. The first North

ENNOBLED BOER GOATS

The American Boer Goat Association (ABGA) recognizes the nation's finest show and breeding goats via its prestigious ennoblement program. To qualify, Boers accumulate show and performance test points that they themselves as well as their sons and daughters earn.

Traditionals (white Boers with red heads) and Non-Traditionals (Boers of any other color) compete in three divisions apiece:

- **Traditional inspected animal seeking Traditional ennoblement.** This Traditional goat must pass two rigorous visual inspections and it and at least three of its similarly inspected Traditional offspring must accumulate 80 ennoblement points earned at ABGA-sanctioned shows and performance trials.

- **Non-Traditional inspected animal seeking Traditional ennoblement.** This Non-Traditional goat must pass two rigorous visual inspections and, while it needn't be shown, at least three of its duly inspected Traditional offspring must accumulate 100 ennoblement points.

- **Non-inspected animal seeking Traditional ennoblement.** One hundred ennoblement points must be earned by at least three of the animal's duly inspected Traditional offspring.

- **Non-Traditional inspected animal seeking Non-Traditional ennoblement.** This Non-Traditional goat must pass two rigorous visual inspections and it and at least three of its similarly inspected Non-Traditional progeny must accumulate 80 ennoblement points.

- **Traditional inspected animal seeking Non-Traditional ennoblement.** This Traditional goat must pass two rigorous visual inspections and, while it needn't be shown, at least three of its duly inspected Non-Traditional progeny must accumulate 100 ennoblement points.

- **Non-inspected animal seeking Non-Traditional ennoblement.** One hundred ennoblement points must be earned by at least three of the animal's duly inspected Non-Traditional offspring.

The best of the best in the Boer world vie for dual ennoblement by earning both Traditional and Non-Traditional titles. As of March 2006, only two goats had earned this highest honor: Powell-Holman Casey and 4 Kids Buttmaster.

When a Boer wins ennoblement, it's noted on his (or her) pedigree and those of all future generations. Boer enthusiasts count the number of ennoblements on their goats' registration papers; the more of the possible 24 ancestors, the better (and the more valuable the goat, especially if its sire and/or dam has been ennobled).

For more information on the American Boer Goat Association Ennoblement Program and to view the Honorees list, see Resources.

Downen R33 "Hoss," champion show buck owned by Matt and Claudia Gurn of MAC Goats in Winona, Missouri, earned the final leg of his ennoblement in May 2006.

American-born Boer goats were released from quarantine in 1993 and sold to breeders throughout Canada and the United States.

An additional influx of South American genetics arrived in 1994 when exotic-animal importer Jurgan Shultz, of the company Camelids of Delaware, together with American goat breeder Norman Kohls and African breeder Tollie Jordaan, selected 400 Boer goats from the best South African herds and had them shipped first to Key West, Florida, then on to Lampasa, Texas, where they were quarantined. At the end of the quarantine period, the goats and their offspring were sold. Animals from this group (and ancestors descending solely from goats imported by Camelids of Delaware) are called CODI/PCI Boer goats (CODI refers to Camelids of Delaware and PCI stands for Pet Center International, the company that handled Shultz's paperwork).

South African goats, meaning fullblood Boers with no lines of descent tracing to New Zealand or Australian genetics (and especially 100 percent CODI/PCI goats), are currently the darlings of many contemporary full-blood breeders. These goats frequently command higher prices than Boer goats descending in part from New Zealand and Australian imports.

Breed Registries

Three American organizations register Boer goats: the American Boer Goat Association, the International Boer Goat Association, and the United States Boer Goat Association. Boer goat registries are unique in that they recognize one another's registration papers — should you buy an International Boer Goat Association–papered goat and prefer to register it with the American Boer Goat Association (or vice versa), you can send the papers to your favorite organization and for a fee, it will record your goat in its herd book and return both sets of papers to you. Likewise, goats registered with any registering body can be shown at any association's sanctioned shows. Nevertheless, registry services as well as each organization's breeding standard (a description of the ideal Boer goat) vary somewhat. Some breeders belong to and register their goats with all three entities, others with one or two. Check them out (they're listed in Resources) to see which one resonates for you.

Description

Boer goats have long, well-muscled bodies. Mature bucks tip the scale at 250 to 350 pounds; does, 150 to 250 pounds. Most Boers are good-natured, docile goats with very little propensity toward flightiness. Does are prolific (twins are the norm, triplets and quads aren't uncommon), fast-maturing (doelings are frequently bred when they're eight to ten months old), good milkers, and excellent mothers. Bucks are also fast-maturing, reaching puberty by six months of age. Boer does have one trait unique to this breed: they have up to four functional teats, making it possible for them to raise triplets with comparative ease. Both sexes breed year-round; does are capable of kidding three times in two years.

Keeping Boers

Because of their size, Boers require more supplementary feeding than many other meat goat breeds. Well-managed Boer kids, however, are easily capable of gaining ½ pound per day (and more) through weaning at three months of age. Boer bucks bred to smaller, thriftier does sire outstanding commercial kids; they're also in demand for producing crossbred breeds like the BoKi and Genemaster (see Composite Breeds, page 37).

Showing

If you want to show your goats, the Boer is your breed. Unlike production-based breeds such as the Kiko, show Boers (fullbloods and percentages alike) are bred to conform to their registries' standards of excellence and can be exhibited at hundreds of sanctioned Boer goat shows held throughout the United States each year. Crème de la crème fullblood and purebred Boers become *ennobled* — the highest honor in the Boer goat world — based on their show wins and those of their offspring (see page 28).

Savannas

At first glance, Savannas look like all-white Boer goats, yet they developed in South Africa as a separate breed. Like Boers, they were developed using indigenous African goats, but under far more hostile environmental conditions, resulting in a tougher, hardier, more disease- and parasite-resistant goat.

Savannas resemble Boers in build, although they aren't nearly as Roman-nosed. Their sleek white coats are nicely accented by black skin, horns, nose, sexual apparatus, and hooves. The Savanna is an exceptionally productive and vigorous breed capable of surviving intense heat and sunshine, but these handsome goats are remarkably cold- and rain-tolerant too.

The first Savannas in North America were imported in 1994 by Jurgan Shultz and quarantined with his CODI/PCI Boer goats. While Savannas are still relatively scarce outside of South Africa, they're gaining in popularity here and in Australia and New Zealand as well. Pedigree International and the

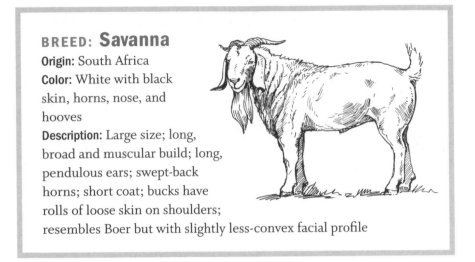

BREED: **Savanna**
Origin: South Africa
Color: White with black skin, horns, nose, and hooves
Description: Large size; long, broad and muscular build; long, pendulous ears; swept-back horns; short coat; bucks have rolls of loose skin on shoulders; resembles Boer but with slightly less-convex facial profile

North American Savanna Association maintain the North American Savanna herd book.

Kikos

If the Boer is the reigning king of the North American meat goat industry, the Kiko is its up-and-coming crown prince. Due to the breed's extreme hardiness and high feed conversation rate (the ability to convert feed efficiently to marketable meat), many goat ranchers who began with Boers are switching to Kikos and Kiko–Boer crosses; Kiko could become America's favorite meat goat breed.

History

The Kiko is a purely production-based breed native to New Zealand. In that country's indigenous Maori tongue, *kiko* means "meat." Kiko goat development began in the 1970s when a group of large-scale farmers who were already farming enormous numbers of feral goats began selecting to produce an extremely hardy meat goat breed capable of not only surviving without intervention under rugged conditions, but also producing fast-growing meat kids in these conditions. Foundation animals weren't assisted at kidding or provided supplementary feed or shelter, and no one dewormed them or trimmed their hooves. The weak died and goats that failed to perform were rigorously culled.

Breed Registries

New Zealand breeders exported a large number of Kikos to the United States in 1994; within two years, the first American registry, the American Kiko Goat Association (the official liaison with New Zealand's Kiko registry), was born. Breeders who tried Kikos in their herds liked them. Bred true or crossed with other meat breeds, especially the Boer, Kikos outdistanced their competition in parts of the United States, especially the far North and in the humid southeastern states where Boers don't always thrive. By 2004, a second Kiko registry, the International Kiko Goat Association, was incorporated. Pedigree International also maintains a Kiko herd book and a roster of Kiko breeders.

Description

Adult Kikos are slightly smaller than most Boer goats of the same age and sex. The Kiko has a strong, straight-profiled head with a bold expression and medium-length, nonpendulous ears. Horns on mature bucks are nothing short of spectacular: enormous, broad, and strong, with a spiraling outward sweep; does are horned as well. The Kiko's muscular, medium-length neck blends smoothly into its long, broad, yet compact body, and well-sprung ribs provide lots of room for the animal's internal organs. Each sturdy, well-placed leg terminates in a tough hoof with a pronounced interdigital division. Kikos

BREED: **Kiko**

Origin: New Zealand
Color: Any color acceptable; much of the North American population is white
Description: Large size; long, broad, meaty build; medium length, nonpendulous ears; straight facial profile; enormous, spiraling horns; short to medium coat

have smooth, supple skin and short- to medium-length coats. Most American-born fullblood Kikos are white, but other colors are fully acceptable.

Keeping Kikos

Kikos are the goat of choice for rough and rugged, low-input goat farming. While lighter-boned than Boers or improved Myotonics (see Myotonics, below), Kikos have large frames that pack on a lot of meat. Engineered for exceptional conversion rates, Kiko kids are born small (to nearly eliminate kidding problems) but grow at an astounding rate. Kikos are exceptional moms, protective and prolific; even first-time kidding does rarely fail to mother their kids. Kiko kids are extremely active, up and seeking a teat within minutes after birth. Kikos are highly efficient browsers requiring little in the way of supplementary feeding. They're also worm-resistant and their hooves rarely need trimming.

Showing

Though the American Kiko Goat Association maintains a conformation standard, there are few opportunities to show Kiko goats. A good way to meet the breed and discuss it with veteran breeders is to attend the International Kiko Goat Association's annual Kikofest; contact the organization (see Resources) for information on upcoming events.

Myotonics

The Myotonic, sometimes called the Tennessee Fainting Goat, is recognized as a heritage breed by the American Livestock Breeds Conservancy, an organization dedicated to preserving and promoting endangered breeds of farm animals and poultry. By marketing wethers and culls for meat and selling your best animals to other conservators as breeding stock, you can turn a profit with goats while helping preserve a piece of our American livestock heritage.

History

The breed's strange odyssey began in the 1880s, when an itinerant laborer named John Tinsley appeared at the door of one J. M. Turner, a farmer in Marshall County, Tennessee. Tinsley, who is believed to have come from Nova Scotia, had in his possession a "sacred cow" (probably a Brahman cow) and four goats that had a peculiar habit of stiffening and toppling over when they were startled. Tinsley worked for Turner for several months, and during that time, a Dr. H. H. Mayberry offered Tinsley $36 for his "fainting goats." Tinsley initially turned him down. Later, he appeared at Dr. Mayberry's farm and accepted the offer. A few weeks later, Tinsley departed with his sacred cow, leaving behind the buck and three does, which formed the nucleus of this old, all-American breed.

Because they were quiet, gentle, and productive and didn't climb or jump fences in the manner of their non-Myotonic kin, the breed became popular with hill-country farmers of the day. In the 1940s, Kentucky and Tennessee breeders shipped Myotonic goats to Texas, where they were selectively bred for greater size and referred to as Texas Wooden Legs.

Both original and improved Myotonics were rediscovered during the 1980s. By selecting for reduced size, breeders soon created a third, smaller type of Myotonic goat for the pet trade. Despite differences in their sizes (Myotonics can weigh from 45 to 200 pounds), all types of Myotonic goats share a subdominant and recessive trait called *myotonia congenita* (see the box below).

MYOTONIA CONGENITA

Myotonia in goats is an inherited neuromuscular condition in which muscle cells experience prolonged contraction whenever a goat is frightened. It affects only muscle fiber (leaving respiration, heartbeat, and other bodily functions intact) and doesn't cause the animal any discomfort. During an episode, a goat's legs and back stiffen and if the animal is off-balance, it topples to the ground. Despite their early moniker, Myotonic goats don't faint; they remain awake and aware throughout such seizures. Some Myotonic goats stiffen quite often, perhaps dozens of times a day; others are affected infrequently. Each held contraction and subsequent relaxation builds muscle mass, explaining why the goats that stiffen most frequently are often the most muscular in any Myotonic goat group. *Myotonia congenita* occurs in other species, too, among them dogs (Chow Chows), horses (American Quarter Horses descended from the famous stallion Impressive), tumbler pigeons, sheep, mice, and humans (in humans it's known as Thomson's Disease).

BREED: **Myotonic**

Also called Tennessee Fainting Goats, Fainters, Texas
Wooden Legs, Nervous Goats,
Stiff-Legs, and Scare Goats.
The Tennessee Meat Goat is
a Myotonic, too, but we'll discuss it in a section of its own.

Origin: United States

Color: Any

Description: Miniature to large size; heavy-boned; massive hindquarters and meaty throughout; straight facial profile; several ear types;
short to long coat; polled or spiraling horns

Keeping Myotonics

Some strains of Myotonic goats are extremely parasite-resistant and all are
highly efficient browsers. They're calm, docile, and nonvocal. Does are easy
kidders and excellent mothers; twins and triplets are the norm. Some (but
not all) strains breed year-round. Both sexes are massively muscled and have
exceptionally strong, dense bones for goats of their size. Naturally polled and
horned varieties exist; Myotonics' horns, when present, resemble those of
Kiko goats. Black and white combinations are the norm, although Myotonics
come in an amazing palette of colors. They can be long- or short-haired.

Apart from their extreme vulnerability to predators, Myotonics' single fail-
ing is that they tend to mature more slowly than other meat goat breeds. The
improved meat-production Myotonic, however, has been specifically selected
for faster weight gains.

Myotonics can be registered with Pedigree International, the Myotonic
Goat Registry, or the International Fainting Goat Registry.

Tennessee Meat Goats

The trademarked Tennessee Meat Goat is a hardy, productive breed developed
by goat breeder/author Suzanne W. Gasparotto at her Onion Creek Ranch in
Lohn, Texas. These animals are descended from big, meaty Myotonic goats
carried from Tennessee to Texas during the 1940s. Ms. Gasparotto created the

breed by selecting the largest and most heavily muscled fullblood Myotonics produced on her ranch and breeding them to others of their kind.

To qualify as a Tennessee Meat Goat, an individual must meet or exceed specific size and muscling parameters, descend from Onion Creek stock, and be visually inspected and certified by a representative of Onion Creek Ranch. Once certified, the animal is eligible for registration in the Tennessee Meat Goat herd book maintained by Pedigree International.

Composite Breeds

In 1995, Suzanne W. Gasparotto began crossing Tennessee Meat Goat bucks with Boer and Boer-crossbred does to create a second trademarked breed, the TexMaster. The resulting composite breed is significantly Myotonic, with just enough Boer to promote faster growth. Like Tennessee Meat Goats, TexMasters are registered by Pedigree International.

Genemasters comprise a composite breed created by crossing Boer and Kiko goats; their herd book is maintained by the American Kiko Goat Association. This breed of goat is created in several steps: First, a fullblood Boer is mated with a fullblood or purebred Kiko goat. The resulting 50 percent Boer/50 percent Kiko (which can be registered as a percentage Boer and a percentage Kiko) is then mated with a fullblood Boer to produce an offspring that is 25 percent Kiko/75 percent Boer. This offspring is in turn mated with a 50 percent Boer/50 percent Kiko goat, and the resulting ⅜ Kiko/⅝ Boer is a registerable Genemaster.

BREED: Tennessee Meat Goat
Origin: United States
Color: Any
Description: Large size; broad, meaty build with massive muscling; medium-length, nonpendulous ears; straight facial profile; spiraling horns; short to medium coat

Composite breeds include TexMasters (Myotonic and Boer) like the buck on the left and his BoKi (Boer and Kiko) counterpart.

The International Kiko Goat Association maintains herd books for three Boer/Kiko composite breeds. An American BoKi is a cross between a registered 15/16 (or better) Boer and a registered 15/16 (or better) Kiko. The American MeatMaker, predominately Kiko, is produced by crossing a registered 15/16 (or better) Kiko with a registered BoKi. The International MeatMaker, largely Boer, is created by breeding a registered 15/16 (or better) Boer with a BoKi.

Kiko/Boer composite breeds do exceptionally well in parts of the United States that aren't suited for Boer goat production, particularly the humid southeastern states. In a commercial setting, these hybrids frequently outperform purebreds of either breed, especially in the field of weight gain. In fact, in the early 1990s, the Goatex Group experimented with Kiko/Boer hybrids and found that hybrid kids reached market weight a full two months earlier than control groups of fullblood Boer and Kiko kids.

Spanish Goats

Spanish goats, also called wood goats (in Florida), hill goats (in Virginia), brush and briar goats (in North and South Carolina), or scrub goats are descended in part from goats brought to the New World by early Spanish explorers.

Their size varies greatly. Spanish bucks weigh in at roughly 80 to 200 pounds and does, 65 to 130 pounds. They come in every conceivable color and in both short- and long-haired versions. Both sexes have impressive, spiraling or swept-back horns. Spanish goats tend to be flighty unless they've been handled since birth.

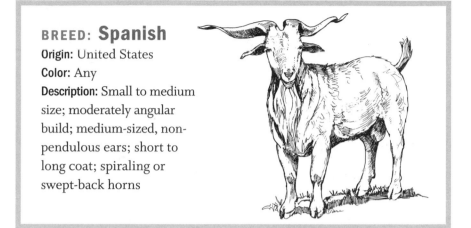

BREED: **Spanish**

Origin: United States

Color: Any

Description: Small to medium size; moderately angular build; medium-sized, non-pendulous ears; short to long coat; spiraling or swept-back horns

In their favor, these goats are tremendously hardy and productive under minimal-input conditions. They're inexpensive and readily available, particularly in the meat goat–producing southern states and Texas. Boer/Spanish crosses form the basis of many commercial herds.

Dairy Goats

When bred to meat-breed bucks, some breeds of dairy does are excellent meat-makers too. Where Spanish goats aren't readily available, commercial producers frequently cross Boer or Kiko bucks with dairy breed does. This works, but with several caveats:

- Dairy does' huge, sometimes pendulous udders are easily damaged under rugged pasture conditions.
- Because they've been bred for extra-long lactations, dairy does are still in full production when their kids reach marketable age. Unless they're milked after their kids are weaned, dairy does often develop mastitis.
- Dairy does require more feed than meat does in order to produce kids and the milk to feed them, making dairy does poor candidates for minimum-input production.

Meat goat producers who use dairy does tend to prefer Nubians. Nubian does are tall (30 inches or higher at the withers), solidly built dairy goats developed in Great Britain during the 1800s and imported to North America as early as 1896. They were once considered a dual-purpose meat and dairy breed. With

When bred to meat-breed bucks, Nubian dairy does are excellent meat-makers, too.

Many producers appreciate the meatiness and fine mothering ability of the LaBoer doe.

their relatively stocky bodies; Roman-nosed profiles; and long, floppy ears, they resemble streamlined Boers, making them ideal candidates for producing percentage Boer show goats.

Other commercial breeders swear by LaMancha dairy does — especially for breeding to Boer bucks to produce LaBoers, odd-looking but meaty crossbreds whose does exhibit fine mothering ability. LaMancha does stand at least 28 inches tall and weigh 130 pounds or more. Though angular, they're compact and wide with well-sprung ribs; a wide chest; a deep, wide barrel; and a wide, long rump. The LaMancha is an earless breed, however, with a strong tendency to pass on earlessness to its offspring. If you don't want earless kids, this definitely isn't your goat.

For additional information about dairy goat breeds, see appendix E, A Milk Goat for the Kids.

Angoras

Angora does don't carry much meat, but especially in Texas and other areas of the Southwest, where these fiber goats are common and relatively inexpensive, Angoras are sometimes used for producing crossbred meat goat kids.

Angoras are small and not very prolific, however. In addition, they require twice-annual clipping to remove their long, curly fiber, and the breed doesn't thrive in cold, damp climates. That said, in *Meat Goats: Their History, Management and Diseases*, Stephanie and Allison Mitcham advocate the use of Angoras in commercial meat goat herds, asserting that the meat of Angora crosses tastes better than that of Boer cross kids. If you're thinking of using Angoras in your program, be sure to read their book.

Registered, Fullblood, Purebred, Percentage, or Crossbred?

When raising goats for meat, you needn't begin with expensive registered goats. If you do buy them, however, it's important to understand what you're getting. Registered goats might be fullbloods, purebreds, or percentages, and some are mixtures of two breeds. Here's how to distinguish them.

Registered Goats

Registered goats are listed in the herd book of one or more breed registries. They sell with registration papers, a certificate or certificates detailing an animal's ownership record, birth date, birth status (single, twin, triplet), color, and identification specifics such as tattoos and ear tag numbers, along with any additional pertinent information. Registered animals are more expensive than unregistered goats of equal quality, but to offset this, their breeding and show stock offspring sell for comparatively higher prices too. Another important reason to choose registered stock: Only papered goats can be shown in breed-sanctioned goat shows.

Listing in the herd book of more than one registry is not uncommon. For instance, a fullblood Boer goat might be registered with the American Boer Goat Association, the International Boer Goat Association, or the United States Boer Goat Association, or perhaps with two or all three of these groups. A crossbred half-Boer and half-Kiko goat with two registered parents could be registered as a 50 percent goat with the three Boer registries and as a BoKi with the International Kiko Goat Association.

When buying a registered goat, carefully compare the information on the registration certificate (such as color, markings, and ear tattoos) with the actual goat to make certain they match. Also check to make sure the last recorded owner is the person from whom you're purchasing the goat; the last recorded owner is the only person who can legally transfer ownership into your name. Some registries issue separate transfer forms, while for others, transfer information is recorded on the registration paper itself. Whichever the case may be, make certain the seller signs the transfer and, if you purchase a pregnant doe, provides a *breeder's memo* (a separate document you'll need to register the offspring of a registered female goat). We'll talk more about paperwork in chapter 5.

Fullblood, Percentage, and Purebred

A fullblood (not fullblooded) goat descends, in full, from ancestors of the same breed. Registered fullbloods generally command the highest prices, and the only way to raise fullbloods is to start with fullblood stock.

Percentage goats are part one breed and part another. Because they generally cost less than fullbloods, they're an economical way to break into raising registered goats. For example, the American Boer Goat Association registers ½, ¾, and ⅞ Boer does in its percentage registry. For their offspring to be eligible for registration, percentage does must always be bred to registered fullblood or purebred bucks. One-half, ¾, ⅞, and $^{15}/_{16}$ Boer bucks aren't eligible for registration, although they do qualify for American Boer Goat Association Record of Pedigree certification.

Fifteen-sixteenths Boer does and $^{31}/_{32}$ bucks are considered American Boer purebreds. These are shown in the same classes as fullbloods and can be used to sire or produce registered percentages, but they're rarely as expensive as fullbloods. No matter how many more generations they're "bred pure," purebred Boer offspring never become eligible for fullblood registration.

Some registries, such as the International Kiko Goat Association, presently allow upgrading to fullblood status; at the $^{15}/_{16}$ level, percentage Kikos become American Fullbloods (and all of the Kiko registries register percentage bucks).

Because percentage goat terminology and registration policies vary from breed to breed, the best way to understand your breed's policies is to contact appropriate registries and ask. One aspect, however, is certain: Fullblood Boers, New Zealand fullblood Kikos, and their counterparts in other breeds generally cost more than animals descended from goats upgraded to purebred or American status.

Crossbreds

Crossbred goats have parents from two separate breeds. These animals benefit from hybrid vigor and are frequently the fastest-growing kids on the block. Some crosses, in fact, are considered breeds in their own right, among them

Genemasters (an American Kiko Goat Association Boer and Kiko composite breed), American BoKis (International Kiko Goat Association Boer and Kiko crosses; see Composite Breeds, above), Meatmakers (International Kiko Goat Association BoKi and Kiko crosses), and Suzanne Gasparotto's unique, trademarked TexMasters (Tennessee Meat Goat and Boer crossbreds). See Composite Breeds, on page 37, for more on all of these crosses.

Producers Profile

John and Sue Weaver
Ozark Goat Trek
Mammoth Spring, Arkansas

John and Sue Weaver live on a ridgetop farm in the southern Ozarks, a few miles south of the Missouri-Arkansas state line. Sue is a full-time author who has written several books about goats. John works with a developmentally challenged client in his home. Due to changing interests and time constraints, the Weavers no longer raise Boer breeding stock, but are glad to field questions via email. They are currently developing a string of Boer pack goats (including several of their former breeding does) for an outfitting venture in the Ozarks.

Q *Sue, how did you get involved with raising meat goats?*

A When I was writing my first book about goats, we visited Matt and Claudia Gurn's MAC Goats in Winona, Missouri, and I fell in love with their old herd sire, MAC Goats Chief Forty-Five, also known as Chiefee. One thing led to another, and because I wanted my own Chiefee, we bought a weanling son of the Gurns' ennobled herd sire, Downed R33 "Hoss," out of one of Chiefee's best-producing daughters. Then we had a registered buck and no Boer does, so we bought some.

Q *What facet of the meat goat industry did you serve?*

A We bred show-quality breeding stock but my writing schedule at the time and caring for all these animals — we raise Miniature/Classic Cheviot sheep and have a menagerie of other livestock ranging from horses to a water buffalo — made it almost impossible to get away to show. Getting your animals out in front of the public is an integral part of breeding show stock. As much as we enjoyed raising show-quality Boers, we didn't feel we could do it properly, so we've phased out our breeding stock over the past few years.

Q What's your take on the future of the meat goat industry? Do you feel there's room in it for new producers?

A Oh my, yes. We came to the Ozarks about eight years ago and the growth of the meat goat industry in these hills is simply phenomenal! This rolling, rocky land doesn't grow much except rough pasture and browse, so this region is essentially goat heaven. People are switching from cattle to meat goats and hair sheep in droves. There's a large, active, USDA-graded sheep- and goat-buying station at the BUB Ranch near Koshkenong, Missouri, and all of the sale barns around here hold huge monthly sheep and goat auctions. The buyers who attend these sales say they still can't get enough goats. The slaughter market in this region is wide open.

Q Do you have any advice for prospective goat raisers?

A Yes. Make sure you have goat- and predator-proof fencing in place before you take on goats; you are sure to be frustrated if you don't. And there is a huge learning curve with any kind of goat-keeping, so start small and learn the ropes before you expand. Buy your goats from responsible, honest sellers. We were extremely fortunate to find the Gurns when we got into Boers and Emily Dixon of Ozark Jewels (who also raises Boers) when we bought our Nubian dairy stock. However, we had a few unpleasant surprises when buying goats from other folks. Assume nothing: Check those mouths, count those teats. It's the only sure way to know what you're getting.

4

Where (and Where Not) to Buy Goats

YOU'VE DONE YOUR HOMEWORK and you're ready to buy goats. The most important thing is to find healthy, productive animals and a breeder who will stand behind them. But where do you buy them? And from whom?

Individual Producers and Breeders

If you plan to raise market goats, you'll probably buy foundation stock from established commercial producers. Unless you start with hundreds of does, you can probably buy them locally, and this is a good idea. Healthy goats already acclimated to your region and spared the stress of long-distance travel stay healthier than goats trucked in from afar.

If you plan to breed high-quality show and breeding stock, however, you'll have fewer goats to select from, and unless you live in a goat-dense state, you won't have many local options. You'll probably have to buy from afar, perhaps through a production sale or on the Internet.

Who Sells Great Goats?

Whichever breed of goat you choose, it's important to locate reputable sellers of healthy stock — but how to find them?

■ **Contact breed registries or the American Meat Goat Association.** Visit their websites and peruse member-breeder directories. Phone or e-mail organizations for additional information.

■ **Look for online breeder directories.** Some reputable ones are listed in Resources. To find more, type *goat directory* in the search box of your favorite search engine (such as www.google.com); qualify it with a breed name (*Boer goat directory*), if you like.

■ **Check out online goat classifieds on websites.** Use your trusty search engine to find them (searches for *goat ads* and *goat classifieds* work well).

■ **Look at ads in goat magazines.** Breed-specific journals such as *The Boer Goat* (published by the American Boer Goat Association) and all-breed publications such as *Goat Rancher*, *Meat Goat Monthly News*, and *The Goat Magazine* are packed with display ads, directories, and classifieds. Subscribe to your favorites or pick them up at major newsstands and farm stores such as Tractor Supply Company (TSC).

■ **Visit breeders' websites.** Type your breed, *goats*, and *sale* into your search engine's search box (that is, *Boer goats sale* or *Kiko goats sale,* for example). Qualify it, if you like, by state *(Boer goats sale Ohio)*. If breeders' websites don't offer what you're looking for, e-mail them directly and ask if they have it. Fullblood breeders might not list percentage and commercial goats online, but they often have them for sale — and if they don't have what you want, they may know someone who does.

■ **Join goat-related e-mail groups.** Breed and general-interest goat lists host "Friday sales" when subscribers can post whatever they want to sell. It's a great way to find all sorts of goats and goat-related supplies.

■ **Take in a goat show.** Visit information booths and chat with exhibitors between classes; it's a best bet for solid information if raising show goats or fullblood breeding stock is your future game. All state and most county fairs host goat shows. Breed associations sanction them too; e-mail or call appropriate organizations for specifics.

■ **Don't overlook local resources.** Look for **GOATS FOR SALE** notices on bulletin boards at feed stores and veterinarian practices; post **GOATS WANTED** notices of your own in these locales. Monitor local classified ads. Talk to vets and County Extension agents in your buying area; they're sure to know who's raising goats near your home. Fiber, dairy, and Pygmy goat breeders often have friends who raise meat goats, so ask them for recommendations.

Which Seller?

Whether dealing with a commercial producer in the next county or a show stock breeder three states distant, you want to assess a seller's reputation before you buy. Simply because he or she markets plenty of goats or runs flashy full-page ads in goat magazines doesn't mean a supplier will sell you healthy, productive animals or stand behind them after the sale.

Fortunately, in this day of Internet commerce, the unscrupulous seller who frequently fleeces buyers has little hope of sweeping his shady dealings under the rug. Duped buyers tend to go public and they're likely to do it via industry-related e-mail lists. Satisfied customers sing sellers' praises via the same forums. These lists are an excellent tool, so use them.

Likewise, before buying from local sellers, tap into the local goat grapevine. Talk to other goat owners in your area. Ask which sellers they would buy from, which they'd avoid, and why.

Once you've narrowed the field to a handful of producers selling your breed of goats, contact them and arrange to visit their farms.

Be courteous and arrive at the designated time. If you already have goats or sheep at home and the prospective seller wants to sanitize your shoes, don't be offended. In fact, consider biosecurity precautions a plus.

Once there, take a look around. Goat farms aren't generally showplaces but they shouldn't be trash dumps either. Are the goats housed in safe, reasonably clean facilities? Are there droppings in the water tanks or are goats eating hay from the ground? Are the goats in good flesh, neither snake thin nor overly fat? In large herds, you're apt to spot a few goats that are skinnier or fatter than the norm, but the majority should be in good working condition.

Ask about a seller's management practices. If you plan to run a minimum-maintenance herd, you don't want to begin with pampered, lot-raised goats. Ask about his or her vaccination and worming philosophies: Which vaccines and dewormers does he or she use and why? How often are the goats vaccinated and dewormed?

Does the seller test for caprine arthritis encephalitis (CAE), caseous lymphadenitis (CL), and Johne's (pronounced YO-neez) disease (see chapter 10 for specific information about these illnesses) and does he or she have documentation? Are any of the goats currently infected? What about foot rot? Soremouth? If these problems have presented themselves in the herd, what has he or she done to control them? Are you shown only the goats you arranged to see or are you able to check out the entire herd? Insist on the latter; if there are problems out there, you want to know about them before you buy.

INTERNET GOAT BUYING

Today, buying goats via the Internet is a way of life. Purchasers prefer Internet sales for numerous reasons. They can:

- Shop for goats anywhere in the world, at any time of day, seven days a week, from the comfort of their own homes.
- Quickly select from a vast pool of animals and breeders.
- Research animals and sellers before they deal, thus saving time and money that would be spent on farm visits.

When buying via the Internet, you should ask the same sorts of questions you would during an in-person farm visit. Of course, if a farm visit can be arranged after you've established genuine interest, so much the better. Here are some tips on pursuing purchasing animals from Internet sellers:

- Be sure you're dealing with reputable sellers. Request sellers' references and check them out; ask others in the industry for constructive feedback on sellers.
- Request video footage of the goats that interest you. If it's not available, ask to see additional photos taken from different angles. Examine this material closely and address any issues before you buy.
- Try to get a written guarantee.
- Be very clear on how you will get the goats home before you make a deposit. Who pays for interstate health papers and the sometimes pricey tests they entail? How long will the seller hold the goats once payment is made and you're lining up transportation? Who foots any vet bills incurred during that wait? What happens if one or more of the goats dies?
- Most important: Get all answers and specifics in writing; don't leave anything to chance.

Ask why these goats are for sale. Is the seller changing bloodlines? Liqui-dating percentage stock in favor of fullbloods? Downsizing? If they're culls, perhaps the trait he or she is culling for doesn't matter to you. For instance, large producers usually don't care to raise bottle babies and often cull triplet and quad producers, while other breeders select for high multiple birthing. Perhaps the goat whose personality drives the seller up a rope is just the sort of doe you'd love to own.

If you like what you see (we'll discuss the specifics of evaluating goats in chapter 5), ask to examine the goats' registration papers as well as their health, vaccination, deworming, and production records. These records should indi-cate a goat's birth status (single, twin, triplet, etc.) and particulars detailing its reproductive career. In addition, ask specifically about each doe's kidding habits. Has she had any birthing problems? Is she a good mom? Records show-ing her kids' birth and weaning weights are especially helpful; a doe who consistently weans about her weight in kids is worth her weight in gold!

Ask about guarantees. Some producers give them, some don't; if there is one and you buy, get it in writing.

Is the seller willing to work with you after the purchase, should questions or problems arise? Ask up front to be certain.

Finally, trust your intuition. If at any time a seller seems evasive or oth-erwise makes you feel uneasy, thank him for his time and look elsewhere. There are too many honest producers out there to deal with someone you don't quite trust.

Production and Dispersal Sales

Production and dispersal sales are ideal venues for buying good goats at great prices. The best sales host an animal preview the day or evening before the event and a complimentary lunch during or just before the sale. They're pub-licized well before sale day and printed catalogs highlight sale goats' pedi-grees and production records. The best way to find out about production and dispersal sales is to peruse breed journals and goat periodicals, such as *The Goat Rancher*. These sales are goat-world social events and can be outstanding places to meet people and purchase quality goats.

Payment in full is expected on sale day. Goats sell with registration papers, health certificates, and any other documentation needed for interstate ship-ment. Guarantees, if any, are stated in the sale catalog. Some sales, especially in the Boer goat world, are televised live over the Internet and preregistered bidders can bid via long distance. The originator of this distance-bidding concept is the DV Auction Company of Lincoln, Nebraska (www.dvauction. com); visit its website to see how it's done.

Nevertheless, it's best to be at such a sale in person and well ahead of time, to perform a hands-on inspection and evaluation of the goats on which you might bid. Study the catalog to see which other lots were consigned by their owners and give those goats the once-over too, checking for suspicious lumps, big knees, or tender hooves that might indicate the seller has CL, CAE, or foot rot present in his or her herd.

Another sensible strategy: Mark your catalog, designating which animals you plan to bid on, and make a notation of your absolute top bid for each. The situation can get exciting when bidding runs hot and heavy, and you don't want to grossly overspend in the heat of the moment.

It's important that if you bring home goats from *any* sale, no matter what their health and vaccination records indicate, plan to quarantine them away from your existing herd. House newcomers in an easy-to-sanitize area at least 50 feet from any other goats. Deworm them, vaccinate them, trim their hooves, and keep them isolated for at least three weeks. Don't forget to sanitize the conveyance in which you hauled them home. For such sanitizing, use one part household bleach to five parts plain water. You can use this solution in a fine-mist spray bottle to thoroughly spray boots and shoes. Be sure to launder all other clothing in hot water and detergent. During the quarantine time, feed and care for your other goats first, so you can scrub up after handling the sale-barn additions. Never go directly from the new group to the rest of your herd. If you can prevent it, don't allow dogs, cats, poultry, or other livestock to travel between one group and the other. When the new goats' time in quarantine is up, sanitize the isolation area and any equipment you've used for them.

Sale Barns

The first rule of goat buying is to buy from individuals or at well-run dispersal and production sales rather than from neighborhood sale barns. Avoid everyday livestock auctions as you would a toothache; they're dumping grounds for other breeders' sick goats and culls.

If you buy at sales barn auctions, you won't know if the goat you choose has been vaccinated, if she's pregnant and by what sort of buck, or what's going on in her herd of origin. She could be an escape artist extraordinaire or a doe who gives birth, then abandons her new kids to their fate. A buck may be infertile — or so dangerous that his owner is willing to see the last of him at any price.

The goat you buy might have been exposed to or be infected with progressively degenerative, slow-incubating diseases such as CL, CAE, and Johne's. Its herd mates could have foot rot or pneumonia. These serious maladies can infect your herd, so err on the side of caution.

Goats at such sale barns that haven't been exposed to disease before they're sold through auction will likely be by day's end. Coughing goats, goats with diarrhea, limping goats, goats dropping soremouth scabs wherever they go — you'll find all of them at sale barns. You definitely don't want to bring these goats (or their pen mates) into your herd.

If you attend such sales, even just to look, you're bound to be tempted to buy. Whether or not you do, scrub your hands using plenty of soap and sanitize the clothes you wore to the sale before going near your goats at home. Use a one-part-bleach to five-parts-water solution to clean your boots and shoes, and remember to launder your clothing in hot water with plenty of detergent. This sounds like overkill, but it isn't. Foot rot, soremouth, respiratory diseases, and a host of other caprine nasties can hitchhike home on your hands and your clothes, so it's best never to take chances.

Finally, as for all other purchased goats, never, ever turn out newly purchased sale barn goats with the rest of the herd. Without exception, isolate them for at least three weeks after bringing them home.

Producers Profile

Janice and Roger Nodine, Southern Belle Farms
Portland, Tennessee

For the past nine years, Janice and Roger Nodine have been raising championship-quality Boers at their Southern Belle Farms, just south of the Kentucky–Tennessee state line. Southern Belle goats consistently win at shows around the country each year. Many are the get and grand-get of EGGSplative Deleted, a massive, red-headed buck who has gone on to greener pastures but who helped put their operation on the map. Janice is a regular contributor to online goat lists and is a respected member of the show goat community.

Q *Janice, how did you get into goats?*

A My father-in-law had a goat that he kept to eat brush down, since he was no longer able to use a Weed Eater at the end of his yard. Well, some dogs killed that goat one night so we bought him another one. Then we built a small pasture and got several more. They were "just goats," but on Sundays, when we ate dinner with his family, we'd go out and feed the goats a little grain and hay. And when Roger and I bought his dad's seven acres, we bought the goats too.

I bred a couple of the does to our neighbor's three-quarter Boer buck and only two caught. But I wanted this type of goat because I thought that buck was beautiful, so we got on the Internet and found some for sale.

Our does were very old, but we wanted those kids, so we bought a three-quarter Boer buck to breed them with. We were so excited — we were going into the meat goat business! But the buck didn't give us what we wanted; we still had plain old goats.

One day we went to an auction and heard the man beside us talking about Boer goats. I almost broke Roger's arm trying to get him to talk to that man. He finally did, and we bought our first fullblood Boer doe. She was one of the first Boer goats to come to the United States. She was heavy bred when we bought her and she had quads when she kidded. We kept a fullblood Boer buck from her and bought some percentage does and a few more fullblood Boer does, and that's how our adventure started.

Now, goats are fun, smart, and very loving critters, so the first time I sold a goat for meat, I knew I didn't want to do much of that. So we decided to start showing and selling breeding stock. It took us nine years to get to where we are today.

Q *Is the meat goat industry strong where you live?*

A Yes! The breeding stock business is very good; we've sent Boer goats from coast to coast and as far north as Idaho. And the meat goat business is booming here in Tennessee and Kentucky. There's a facility in Bowling Green, Kentucky, that outgrew its old sale barn because it couldn't handle the number of meat goats coming through. The new facility is designed specifically for meat goat sales, although they handle other animals too. This place is giving great prices, and they have the best meat goat buyers coming in. They grade the goats so the meat goat has to be better than it used to be. We don't see many scrawny, sickly goats going through that sale because the grading system makes producers look at what they're trying to sell. It makes goat ranchers take a little extra care in raising their stock so they'll grade out as prime.

Q *Do you think there's room in the industry for more goat producers?*

A Oh, the goat industry is just getting started and there's lots of room for more producers. There are billions of pounds of goat meat still being imported and it's time the American people get in on the meat goat industry and do well with it.

Q *I was impressed with something you said about buying herd sires. Could you repeat that for our readers?*

A Well, a lot of people think if you buy a kid from show-winning parents, you're bound to have a show goat and a good herd sire. But that's not necessarily true. We kept four buck kids out of our kid crop this year and sold the rest for meat. The four we kept don't seem to be show prospects but they have everything they need to be good herd sires: length of loin, good topline, good mouth, they're up on their pasterns, they track well, they're Roman-nosed and have good teat structure, good hips, good scrotum — some of these bucks look terrific, but they wouldn't do well in the show ring because they just don't have everything that you need to see in show bucks. Showing, especially bucks, is very demanding and you have very stiff competition out there. But we think these four bucks will make good herd sires, and herd sires are the heart of a goat business.

We bought a lot of young herd sires ourselves, and this last time we got it right. ED [EGGsplative Deleted] improved our genetics one hundred times over and we've never regretted buying him or any of the other bucks that we used along the way. We go by pedigrees as well as by a buck's structure, and then we have to breed and wait to see the results.

Q *Do you have any advice for people thinking of entering the meat goat business?*

A Yes: Before buying any type of goat, visit farms — not auctions — and look at goats. Inquire about their health and ask what they're vaccinated for and how they are kept. And even after you get goats, keep on educating yourself and browsing around. The best people I've met have been through goats and I've made many friends. The goat world is a friendly place and we're real glad we're part of it.

5

Selecting Breeding Stock

IT'S IMPORTANT TO KNOW WHICH FACET of the meat goat industry you're going to serve before you purchase goats. If supplying commercial slaughter kids to the ethnic meat market is your pleasure, you won't need stock that meets specific breed standards; if you plan to breed show and high-quality, registered fullblood breeding stock, you certainly do.

Picture this: You're in the market for a new herd sire. The one you favor is a long, meaty, fullblood Boer buck. He's well bred, with plenty of ennoblements in his pedigree; he's impressively built; and he's priced to sell. His color isn't quite right, however. He's mostly white with a splash of red on one side of his face, one of his eyes is blue, and the skin under his tail is mostly pink. This boy won't win ribbons (under-colored, under-pigmented, and blue-eyed show Boers just don't fly), but he could be an outstanding sire of fast-gaining, meaty kids in a commercial setting. Is he a bargain? Only if he truly fits your needs.

Another example: Producers of winning 4-H and FFA show wethers usually use a different sort of fullblood Boer buck than breeders of standard show ring Boer goats. To serve the top-end show wether market, you'll need a long, tighter-skinned, somewhat leggy buck. Short-legged, mega-muscular, loose-skinned bucks don't sire today's winning wethers.

It can't be stressed enough: To avoid costly false starts, study your market before you go shopping.

Don't Buy Trouble

Before you buy foundation stock, learn to recognize healthy goats. All goats are prone to serious diseases like CAE (caprine arthritis encephalitis) and Johne's disease, which can put you out of business. If other illnesses such as CL (caseous lymphadenitis) gain a toehold in your herd, knowledgeable buyers will refuse to buy your stock. Some illnesses such as pneumonia and other infectious respiratory diseases can rip through your herd, kill goats, and run

HEALTHY GOAT, SICK GOAT

Healthy Goats	Sick Goats
Are alert and curious.	Are dull and uninterested in their surroundings; frequently isolate themselves from the rest of the herd.
Have bright, clear eyes.	Have dull, depressed-looking eyes; may have fresh or crusty opaque discharge in the corners of their eyes.
Have dry, cool noses (a trace of clear nasal discharge isn't cause for concern); breathing is regular and unlabored.	May have thick, opaque, creamy white, yellow, or greenish nasal discharge; may wheeze, cough, or breathe heavily and/or erratically.
Have clean, glossy hair coats and pliable, vermin- and eruption-free skin.	Have dull, dry hair coats; skin may show evidence of external parasites or skin disease.
Move freely and easily.	Move slowly, unevenly, or with a limp.
Have average weight for their breed and age.	May be thin or emaciated. (Goats with CAE, scrapie, and Johne's disease become increasingly emaciated as these fatal diseases progress — never buy sick-looking, skinny goats!)
Have healthy appetites (most goats love food and will mob you for it).	Won't eat.
Ruminate (chew cud) after eating.	Don't chew their cud.
Have firm, berrylike droppings; tail and surrounding area are clean.	May have scours (diarrhea); tail area and hair on hind legs may be matted with fresh or dried diarrhea.
Have normal temperatures (101.5–104.5 for adult goats; slightly higher temperatures are acceptable in kids).	Run high or low temperatures. (Subnormal temperatures are generally more worrisome than fevers.)

up astronomical bills with your vet. Avoid unnecessary frustration, hassles, and monetary loss; do your level best to avoid buying sick goats.

Testing for CAE and Johne's is routine in the dairy goat world, but most meat goat breeders haven't yet climbed aboard that train. Insist that expensive purchases be tested for these progressively fatal diseases, or buy from breeders who do test their herds. Ask to see documentation; when paying a premium for tested goats, make certain you get what you're paying for.

An increasing number of meat goat producers are testing their stock for CL. Still, carefully examine potential purchases, searching for abscesses and suspicious bumps or scars, especially in the throat, neck, and groin areas. Not every abscess is caused by CL (inoculation site nodules, for instance, are very common), so if you find a goat you like but it has a suspicious lump, ask to have the lump drained and its contents cultured at your expense. If the seller isn't willing to do this, it's probably best not to buy his or her goats.

Breeding Better Boers

If you plan to show, you'll likely raise Boer goats. Other breeds can be shown, but not as extensively given that their registering bodies don't sanction shows throughout the country, as Boer goat registries do. At this writing, there are three Boer goat associations, and each organization's standard is slightly different from the rest. All North American standards, including that of the Canadian Boer Goat Association, are based on the standard written by the Boer Goat Breeders Association of South Africa.

The ennobled buck (left) and doe (right) shown here conform to the South African standard.

Those who raise other meat goat breeds, take note: Apart from color and certain head, horn, and ear parameters, this standard can be applied to other productive meat goat breeds too.

Breed Standards of the South African Boer Goat

The following are the breed standards of the Boer Breeders Association of South Africa. The aim of breeding standards is to improve the breed and increase a goat's economic value.

Conformation

Head: A strong head with large, soft brown eyes and without an untamed look. A strong slightly curved nose; wide nostrils; strong, well-formed mouth with well-fitted jaws. Up to 4-tooth-olds must show a 100 percent fit. Six-tooth-olds and older may show 6 mm protrusion. Permanent teeth must cut in the correct anatomical place. The forehead must be prominently curved, linking up with the curve of nose and horns. Horns should be strong, of moderate length, and placed moderately apart with a gradual backward curve. Horns must be as round and solid as possible and colored darkly. Ears are to be broad, smooth, and of medium length, hanging downward from the head.

Characteristic cull defects: Concave forehead; horns too straight or too flat; pointed jaw; ears folded (lengthwise); stiff, protruding ears; too-short ears; too-long lower jaws; short bottom jaw; and blue eyes.

Neck and Forequarters: A neck of moderate length in proportion to the length of the body, full and well-fleshed and well-joined with the forequarter is essential. A broad breastbone with a deep and broad brisket. The shoulder should be fleshy, in proportion to the body, and be well-fitted to the withers. The withers should be broad (not sharp). The front legs should be of medium length and in proportion to the depth of the body. The legs should be strong and well-placed, with strong pastern joints and dark, well-formed hoofs.

Characteristic cull defects: Too-long, thin neck; too-short neck; shoulders too loose.

Barrel: The ideal is a long, deep, broad barrel. The ribs must be well-sprung and fleshed, and the loins as well-fitted as possible. The goat should have a broad, fairly straight back and must not be pinched behind the shoulders.

Characteristic cull defects: Back too concave, too slabsided, too cylindrical or pinched behind the shoulder.

Hindquarters: A broad and long rump, not sloping too much, with well-fleshed buttocks and thighs. The tail must be straight where it grows out of the dock and then swing to either side.

Characteristic cull defects: A rump that hangs too much or is too short, a too-long shank, or flat buttocks.

Legs: The legs must be strong (of good texture) and well placed. Strong legs imply hardiness and a strong constitution, which are absolutely essential characteristics of the Boer Goat.

Characteristic cull defects: Knock knees, bandy legs, "koeisekel" (too straight in the hock), or "regophak" (sickle leg); legs too thin or too fleshy; weak pasterns; hoofs pointing outward or inward.

Skin and Coverings: A loose, supple skin with sufficient chest and neck folds, especially in the case of rams. Eyelids and hairless parts must be pigmented. The hairless skin under the tail should have 75 percent pigmentation for stud purposes, with 100 percent pigmentation the ideal. Short, glossy hair; limited amount of cashmere (fluffy undercoat) permissible during winter months.

WHAT ABOUT CULLS?

Some producers take culled goats (even papered fullbloods) to the local sales barn; others sell to breeders to whom an animal's faults aren't a major issue. For instance, because we enjoy feeding bottle babies, triplets are a blessing at our farm, and a goat that can't stand steamy Louisiana summers might do well for you in Illinois.

Keep this in mind when buying your goats: Someone else's "trash" might be just what you need!

When asked why they cull goats from their herds, subscribers to several major goat listservs gave these reasons.

REASONS DOES ARE CULLED

- They have poor mothering skills (especially does that abandon their kids); allowances are usually made for first-time kidders.
- They're poor keepers (thin, unhealthy).
- They have poor conformation that's passed on to offspring.
- They have bad mouths (overbite, underbite).
- They're unable to handle temperature extremes (heat or cold).
- They have bad udders (fish teats, cluster teats, more than four teats, teats too large for newborns to nurse; extremely pendulous udders).

Characteristic cull defects: Covering too long and coarse or too furry.

Sexual Organs: Ewes* — Well-formed udder firmly attached with no more than two functional teats on a side. Permissible defects: If a teat is not separated, but there are two milk openings; for double teats, the front 50 percent should be split. Rams — Two reasonably large, well-formed, equal-sized testes in one scrotum. The scrotum must be at least 25 cm in circumference.

Characteristic cull defects: Bunched, calabash, or double teats; too-small testes, a scrotum with more than a 5 cm split.

Quality: This is achieved with short, glossy hair and a fine luster.

Size: The ideal is an average-sized, heavy goat with maximum meat production. A desirable relationship between length of leg and depth of body should be achieved at all ages. Lambs should tend to be longer in the leg.

*In South Africa, sheep terms are used to denote the sex of goats.

- They produce little or no milk.
- They fail to conceive two years in a row.
- They consistently produce single kids.
- They birth multiples (three or more kids) several years in a row.
- They're troublemakers that are always getting out, walking down fences, pushing others away from feed, and so on.

REASONS BUCKS ARE CULLED

- They consistently sire unusually large, hard-to-birth kids.
- They sire poorly muscled kids.
- They sire slow-growing kids.
- They're poor keepers (thin, unhealthy).
- They have poor libidos.
- They are overly aggressive toward humans.
- They've sustained injuries to legs or feet that make covering does very difficult or impossible.
- They have split scrotums.
- They have poor teat structure (same criteria as does).
- They have poor conformation that is passed on to offspring.
- They exhibit destructiveness (especially fence smashing).

Characteristic cull defects: Goats too large or too small.

Coloring: The ideal is a white goat with a red head and ears, and fully pigmented. The blaze must be evident. Shadings between light red and dark red are permissible. The minimum requirement for a stud animal is a patch of at least 10 cm in diameter on both sides of the head, ears excluded. Both ears should have at least 75 percent red coloring and the same percentage pigmentation.

The following is permissible for stud purposes:

- Head, Neck, and Forequarters: A total red coloring not farther than the shoulder blade; on the shoulder, not lower than level with the chest junction.
- Barrel, Hindquarters, and Belly: Only one patch not exceeding 10 cm in diameter.
- Legs: Patches with a maximum of 5 cm in diameter below an imaginary line formed by the chest and the underline.
- Tail: The tail may be red, but the red color may not continue onto the body for more than 2.5 cm.
- Red Hair Covering: Very few red hairs at the 2-tooth stage.
- Pigmentation: Discriminate against too-light pigmentation.

Note: South African breeders frown on anything but what we in North America call Traditional coloring. Most American show Boers are Traditionals, but black, red, and paint (spotted) Boers are rapidly gaining favor and are eligible for registration in all three American Boer goat herd books.

Explanation of Breed Standards

Many aspects cannot be fully defined in applying standards, and in such cases, the inspector or judge must use his discretion. In spite of the breed standards being clear and to the point, it is necessary to supply additional information with respect to certain descriptions. The major part of the body of the goat must be white to make it conspicuous and to facilitate the rounding up of goats in dense terrain. A pigmented skin on the hairless parts, e.g., under the tail, around the eyelids and mouth and so forth, is absolutely essential, because it offers resistance to sunburn, which may result in cancer, and is also more resistant to skin disease. An animal with loose, supple skin and short hair is better adapted to the climatic conditions in South Africa. In addition, a skin of this kind provides resistance to external parasites.

General Appearance and Type: This is a goat with a fine head; round horns bent backward; a loose, supple, and pleated skin (especially in rams); with different body parts well-fleshed in perfect balance. The ewe must be feminine, wedging slightly to the front, which is a sign of fertility. The ram appears heavier in the head, neck, and forequarters. The Boer goat is a symmetrical animal, with a strong, vigorous appearance and enough quality. In the ewe there is strong emphasis on femininity; in the ram, one of masculinity.

Fertility: Shows — a ewe must have lambed at six-tooth age already or must visibly be with young or she will be culled. Auctions — six-tooth and older ewes must visibly be with young or be certified in writing as pregnant by a veterinary surgeon or the ewe will be culled.

LAMBS, HOGGETS, AND TWO-TOOTH GOATS

In South Africa (as well as Britain, Australia, and New Zealand), goats (and sheep) are aged according to the number of pairs of fully erupted permanent incisors in their mouths. Some American authorities use this method of aging too.

Lamb: a kid with eight temporary milk teeth

Hogget: a kid with eight temporary milk teeth, with the center pair of permanent incisors beginning to erupt around 12 months of age

Two-tooth: a goat with one center pair of permanent incisors (12 to 18 months)

Four-tooth: a goat with two pairs of permanent incisors (21 to 24 months)

Six-tooth: a goat with three pairs of permanent incisors (30 to 36 months)

Full mouth: a goat with all permanent incisors in place (42 to 48 months and older)

Broken mouth: a goat that is losing its permanent teeth

Gummer: an elderly goat with no teeth

WHERE'S THE MEAT?

Though breed character (defined as how closely an individual measures up to the ideal for its breed) varies from breed to breed, all well-built meat goats share certain traits. The following are starting points. If you plan to raise show goats or replacement breeding stock, however, contact your registry for a copy of its breed standard and know it by heart before you buy.

- When evaluating meat goats, examine their heads. While the elegant, convex profile of a Boer buck is unlike that of straight-profiled Kiko and Myotonic bucks, all should have strong, wide foreheads (generally, the wider the forehead, the wider the goat will be); wide, flaring nostrils; and well-set horns. Avoid goats with narrow-based horns, which can trap another goat's leg and break it.

KIKO BUCK

BOER BUCK

- Goats have a hard pad of tissue called a dental palate instead of upper front teeth. To efficiently browse or graze, a goat's lower teeth must meet the leading edge of its dental palate. Refuse individuals with pronounced "monkey mouth" (underbite, sometimes called "sow mouth"; the lower teeth protrude ahead of the dental palate) or "parrot mouth" (overbite; the dental palate protrudes in front of the lower teeth). Though Boer goats have a tendency to be slightly monkey mouthed, for show purposes, a young goat's teeth must touch the dental pad; an aged Boer's teeth should also be flush or protrude no more than about one-quarter inch.

Sound mouth   Monkey mouth   Parrot mouth

■ Meat goats of all breeds should have broad, muscular forequarters. The longer the goat, the meatier it will be. The best Boer bucks are amazingly long, especially through their loins. To check for length, draw an imaginary box around a meat goat. If it's square, the goat isn't especially lengthy; if it's a long rectangle, there's your better goat. Viewed from front or back, a good meat goat will be broad with a deep chest (brisket) and have plenty of space between its legs. The cuts of meat most buyers prefer are from the loin and hindquarters, so wide, muscular hindquarters are important.

WHERE'S THE MEAT? continued

- Choose goats with plenty of body depth, width, and spring of ribs when viewed from top and sides. (A goat with well-sprung ribs has a flat back and wide sides.) These goats will be meatier than slab-sided goats — and wider, deep-bodied does have plenty of room to carry kids.

- Meat goats should have strong, broad straight backs. Avoid sway-backed young goats but older does, even good ones, tend to show a bit of sway. Does with steep rumps that drop off sharply from hip bones to tail are said to be more prone to kidding problems than their longer-rumped sisters (and long rumps are meatier than short ones).

A doe with a straight back and good legs   A swaybacked doe

- Better meat goats have strong, straight legs set perpendicular to the ground, and the best ones have plenty of bone. Goats should have strong, nearly upright pasterns. Look for tight toes and a level sole. These are important longevity features.

A buck with good legs and a straight back

- Select bucks with large, firm testicles of equal size, and avoid bucks with split scrotums. Does' udders should be free of lumps and well attached and should feature functional, well-spaced teats.

A buck with large, symmetrical testicles

A buck with a split scrotum

A doe with a well-attached, symmetrical, capacious udder with normal-size teats

A doe with a pendulous, lopsided udder and sausage teats

- Don't select for body size alone; ask to see production records before you buy. A small doe that consistently raises several fast-growing kids is better than a large doe that lacks mothering skills or produces slow-maturing singles — and the smaller one will eat much less feed.
- If you plan to market goat leather or pelts as secondary products, choose goats with loose, supple skin and fine, soft hair; these features generally equate with high-quality, fine-grained hides.

Look 'em Over Before You Buy

Carefully examine prospective breeding stock before you buy it. If you can't visit the seller to do it yourself, send a trusted, goat-savvy representative or be absolutely certain you're buying from a scrupulously honest seller.

Choose Healthy Goats

Even if goats come with vet-issued health papers, you need to look over the animals yourself before you take them home. Goats can get sick in a very short time; just because they were well two weeks ago, when the seller's vet checked them, doesn't mean they're still healthy when you arrive to pick them up.

Teeth

Always check the teeth of goats before you buy. An older goat whose bite is slightly off may be acceptable but a kid with bad occlusion, especially if it's a Boer, will get much worse as it matures, and malocclusion is more than a cosmetic issue. Goats with a severe over- or underbite (see page 63) aren't able to graze or browse effectively, so you'll have to supplement their feed to keep them in the pink. What's more, bite is considered hereditary — you don't want it to be passed on to future generations.

It's the norm for an older goat to have missing teeth. Broken-mouthed goats and gummers don't do well in large herd settings, where they have to compete with younger goats for their share of grub, yet show and breeding stock producers seeking high-quality fullbloods at modest prices should consider broken-mouthed oldsters. Many well-managed does and bucks breed well into their early teens; it just takes extra feed and a touch of TLC to keep them going. (See box on page 61.)

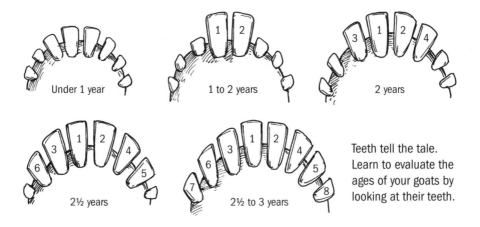

Under 1 year

1 to 2 years

2 years

2½ years

2½ to 3 years

Teeth tell the tale. Learn to evaluate the ages of your goats by looking at their teeth.

Horns

Most meat goats have horns. Know what's normal for your breed and choose accordingly.

Some producers don't want goats with horns, which can get caught in fences and feeding equipment. In addition, horned goats kept in confinement tend to use them on one another, and an occasional crabby goat will jab and butt a pasture mate no matter how much room they share. Too, a horn can easily put out an eye. In most states, goats shown in 4-H must be hornless or have their horns blunted.

If you don't want goats with horns, raise your own kids or buy newborn bottle babies and disbud them. Alternatively, buy kids from a breeder who agrees to disbud them for you. Disbudding consists of applying a red-hot iron to each of a young kid's barely emerging horn buds and holding it there until the horn bud is destroyed. If done in one pass, this takes roughly 10 seconds per horn. Done correctly the process works, but it's not without risk — improperly disbudded, brain-damaged kids can and do die.

Dehorning adult goats is much nastier, however, so if you want hornless goats, it's best to disbud kids while they're tiny. Dehorning is a bloody, traumatic mess to be undertaken only by a vet experienced in the procedure. A goat's horns form a part of its skull. Sawing them off leaves gaping, bloody holes leading directly into its sinus cavities. It requires weeks of painful treatment for them to heal. Sometimes, especially in Boers, an old buck's curling horns grow so close to the sides of his face that he can't chew in a normal sideways motion; these bucks *must* be dehorned. Otherwise, the practice is unnecessarily cruel. If you don't want horned adult goats, don't buy them!

Technically, disbudded and dehorned goats can be shown in Boer goat shows. In reality, however, these goats rarely place at the larger shows. It's hard to determine if a Boer has acceptable horn placement when its horns have been removed.

Horns on meat goats have their uses as well. Because goats hate being dragged around by their horns, and because horns are filled with blood vessels and are easily broken or hurt, don't do it as a matter of course — but in an emergency, those horns make excellent handles if grasped at their base, where they're strongest.

Teats

The South African farmers who developed today's Boer goat selected for does with four functional teats. Such does, they reasoned, were more able to raise multiple kids successfully than does with only two teats. Some American Boer breeders prefer four-teated stock, others are breeding back to two-teatedness. Both sets of breeders, however, want "clean" teats — that is, fully functional, easy-to-nurse teats with single orifices and as few extraneous nubs and bumps as possible.

Unfortunately, many Boer and Boer-cross goats have poor teat structure, so when buying breeding stock, it's important to examine teats (even those of bucks; bucks pass teat structure to their offspring too).

In this illustration, **A, B, E, G,** and **H** have too many teats per side. **B** and **C** are *antler teats.* **D** is a *fish* (or *fish tail*) *teat.* **E** has too many teats, all having orifices (because the fish teat counts as two teats). **F** is a *kalbas (gourd) teat.* **H** is a *cluster teat.* If the small teat on the far left of **G** had no orifice, it would be called a *nub,* and if the longer of the two teats allowed nursing, this would be considered an acceptable arrangement.

The following points indicate acceptable teat structure:

- Four-teated goats' teats should be widely spaced; teats are ideally (but rarely) all the same size.
- Many American breeders prefer two functional teats.
- Small bumps and nubs are acceptable as long as they lack orifices and don't interfere with kids' suckling. Large teatlike nubs (blind teats) should be avoided; newborn kids will try to suckle them and may miss out on their share of life-giving colostrum.
- Fused teats with either one or two orifices are also acceptable, but *only* if they're slender enough for kids to suckle.

The following points indicate possibly acceptable teat structure:

- The American Boer Goat Association considers fish teats such as those shown in the illustration (D) as disqualification factors for showing. A fish teat split for at least 50 percent of its length can be nursed, however, so is acceptable to some goat producers.
- More than two teats per side is a showing disqualification, but if they're widely separated, a kid can nurse from them, thus they're acceptable.

The following point indicates unacceptable teat structure:

- A normal teat is paired with a fish teat on an udder to make three teats per side, a double disqualification in the show ring.

The following points indicate highly unacceptable teat structure (these teats not only result in show disqualification; kids can't nurse them either):

- Antler teats: A kid nurses from one branch of an antler teat while milk from the other orifices is likely to spill onto its face; if the kid aspirates this spilled milk, it could contract pneumonia as a result and die.
- Gourd (kalbas) teats: These are too big for a kid to get in its mouth.
- Cluster teats: These are unacceptable for the same reason as antler teats.

A Few Words about Registration Papers

We touched on this topic in chapter 2, but it bears repeating: Carefully check the registration papers of any goat you're buying to make certain

READING A BOER GOAT PEDIGREE

The abbreviations used in a Boer goat's pedigree can tell you where his ancestors came from. These are a few of the most common.

K: Keri Downs, North Island, New Zealand (Landcorp Farming Limited)

E: Erewell, South Island, New Zealand (Landcorp Farming Limited)

Z: embryos from Zimbabwe; ancestors of the Landcorp goats

WW: African Goat Flock, New Zealand (fullblood Boer)

WG: African Goat Flock, New Zealand (percentage Boer)

BR: ancestors of the fullblood Boers in African Goat Flock's WW herd

A: Australia Breeding Management (Australia)

bbb/nnnn: (South Africa or South African frozen embryos implanted in Canadian recipient does); bbb represents the number assigned to each South African breeder by the South African Boer Breeders Society; nnnn represents an individual animal's identification number within that herd.

O: Olds College (Canada) in partnership with Landcorp Farming Limited (New Zealand)

TR: goats registered with the Canadian Boer Goat Association

you're getting what you're paying for. Some sellers advertise "registered goats." Some sellers don't know (or pretend they don't know) the difference between fullblood, pureblood, and percentage papers, so don't assume registered means fullblood.

Here is a list of important points with regard to registration papers:

- **Transfer of ownership.** Make certain the goat is recorded in the seller's name. The last recorded owner is the only person who can transfer the goat to you. In some cases, the seller has a transfer form made out in his name but has never gotten around to having the papers updated. Some (but not all) registries allow this type of seller to issue a second form transferring impending ownership to you; if you (and your registry) accept this arrangement, be prepared to pay double transfer fees.

- **Service memos.** If you're buying a bred doe, you'll need a service memo signed by both the buck's and the doe's owners in order to register her kids. Depending on the registry you do business with, the service memo may be a separate document or part of the transfer slip. Know which you need in advance.
- **Registering kids.** Most registries stipulate that kids must be registered by their breeders. If you buy eligible but as yet unregistered kids, make certain you get a fully filled out and signed registration application and a completed transfer slip transferring their ownership to you.
- **Pedigrees.** Most registration certificates have a four- or five-generation pedigree printed somewhere on the paper. Breeding stock producers need to be sure they're getting the pedigree they assume they're getting.
- **South African Boer goats.** Boer goats of straight South African breeding, especially CODI/PCI goats, are animals whose ancestors came directly from South Africa to North America; none of them was raised in Australia or New Zealand. These goats sell for premium prices. Every line of CODI/PCI goats can be traced, without exception, to South African goats imported by Jurgan Shultz. Don't pay for straight South African bloodlines unless you buy from a seller with an impeccable reputation or know

PEDIGREE JARGON

Goats are said to be *by* a specific buck and *out of* a certain dam. Thus, MACG Mac Goats P-123 "Kahless" is by Downen R33 "Hoss" Ennobled and out of MACG Mac Goats Droplet.

Some breeders, however, might say MACG Mac Goats P-123 "Kahless" is by Downen R33 "Hoss" Ennobled *over* MACG Mac Goats Droplet. Others might say "MACG Mac Goats T-320 "Kahless" is *out of* DOW Pipeline, meaning DOW Pipeline is one of his ancestors (DOW Pipeline is a paternal great-grandsire of MACG Mac Goats P-123 "Kahless"). In this last instance, if you thought you were buying a son of DOW Pipeline, you'd be sadly mistaken.

pedigrees well enough to judge them for yourself. One exception: If you buy straight South African Boers registered with the International Boer Goat Association, the letters SA will be appended after their registered names. Otherwise, let the buy beware!

Closing the Sale

Before handing over your check, get all applicable guarantees and sales conditions in writing. Do it every time, even when dealing with friends. People misunderstand or forget. A comprehensive, written sales contract is an absolute must, so hammer out these details and write them down!

Producers Profile

Mona and Joey Enderli, Enderli Farms
Baytown, Texas

Though Mona and Joey Enderli no longer breed goats, Mona's advice is still right on. The Enderlis used to raise Boers on their farm, located 30 miles east of Houston on the flat coastal plains of the Gulf of Mexico. Enderli Farms was a multipurpose operation producing both show-quality breeding stock and winning show wethers.

Q *Mona, how did you get into the goat business?*

A We started out very small . . . two mixed-breed, three-month-old kids. We were immediately hooked, and within a week we bought ten more does. Nothing in either of our backgrounds prepared us for goats. Joey has lived in Baytown all of his life, participated in 4-H, and showed cattle. I grew up in a very small farming community in Oklahoma. My grandparents farmed, and I spent every minute I could on that farm. That was the best place on earth for a kid to be.

Prior to moving to our present location, we lived in town and had a farm with BeefMaster cattle out in east Texas. After more than a decade of making the six-hour, round-trip drive almost every weekend and holiday, we bought the farm where we now live and moved our cattle down here so we could enjoy them without the drive. But then Joey saw some baby goats at a neighbor's house where he was baling hay, and the rest is history.

Q *Why did you choose Boers?*

A Joey's mom had Boer goats and talked about them constantly. We decided to check them out and see what was so great about them. We were taken with their regal appearance and sweet disposition — and their value seemed to be increasing steadily while a lot of other breeds were commanding only rock-bottom prices. Adjusting to the moist, humid weather of the Gulf Coast region has been a real challenge for them, but with patience and a watchful eye, they've made the transition and are thriving.

Q *What's your take on the meat goat industry? Is it strong in your locale?*

A The meat goat industry in the Houston area has expanded tremendously in the last five years due in large part to the increasing Muslim, European, and Caribbean populations. For these groups, goat meat is their normal cuisine, and what better way to fill that market niche than with a meat goat such as the Boer?

Q *Do you feel there's still room in meat goats for new producers?*

A People raise goats for a variety of reasons: as pets, for show, for meat or milk, as replacement stock. So whatever reason a person might have, there seems to be a market to fit his or her situation. People entering the industry now have much better information available to them than was available to the earlier breeders. And the Internet has made that information as close as their fingertips.

Q *What advice would you give someone thinking about starting out in meat goats?*

A My advice to anyone considering buying meat goats is to have a business plan and know your markets and to make sure you have the time and energy to commit to the endeavor. This business takes dedication. Raising goats isn't easy. If you're not prepared to work long hours in all kinds of weather, you may want to consider a different business. Be aware that raising goats isn't cheap! You can't scrimp on pasture, feed, housing, or medicines. If you're not prepared to provide your goats with the best care possible, again, you may want to consider another business. What you may save by scrimping will be lost when your goats die from lack of proper care.

This business is twenty-four hours a day, seven days a week, three hundred and sixty-five days a year, and it's not for the faint-hearted. Would you be able to deliver a kid that is breech, too large, or possibly dead in utero? If seeing blood makes you feel faint, you don't like being dirty, the thought of having goat poop stuck to your shoes all the time is disgusting to you, or you expect to be able to take a normal vacation at the drop of a hat, then you'd better rethink your plans.

I don't want to scare anyone away from raising goats, but I do want to make sure you know what you're getting into. There's a huge learning curve, and during the process you'll lose some goats, so don't spend your life savings on your first goat purchases. Wait until you've learned enough to feel confident in what you're doing every day before plunking down your hard-earned money on top-quality genetics.

6

Think Like a Goat

GOATS ARE INQUISITIVE, ACTIVE, AGILE, and intelligent crea-
tures. Ask anyone who keeps goats and he'll tell you the same thing: It's better
to work with goats than to try to work against them.

Fortunately, it's easy to learn to think like a goat. When you understand
what makes goats tick, you can assume leadership of your herd and head off
impending problems before they begin.

Range and Feeding Behavior

Goats are browsers, not grazers. They prefer to range over a large area and
eat a widely varied diet based on weeds, wild herbs, shoots, brush, twigs, and
bark. They won't eat plants contaminated with goat urine or goat droppings,
nor will confined goats eat soiled, moldy, or musty feed.

Goats prefer to nibble the tops of plants rather than devouring them down
to ground level in the manner of grazing species. Since internal parasite
larvae tend to spend their time close to the earth, it's fairly easy to control
worm infestation if goats are allowed to range over the large areas they prefer or they're moved to new pastures before old ones are grazed close to the
ground.

Under range conditions, goats are on the move, feeding for up to 12 hours
per day. They spend their downtime dozing and ruminating. Goats don't
ruminate, however, if they're nervous or on the alert.

When alarmed, goats curl their tails tightly over their backs and emit a
high-pitched, sneezelike sound, sometimes stomping a forefoot as they do so.

Goats are browsers. They nibble the tips of plants and then move along, rather than grazing plants down to the ground. They also stand on their hind legs to reach morsels like shrub brush and leaves.

Alarmed goats flee a short distance and then turn to face whatever frightened them. If pursued, they often scatter.

When particularly annoyed or angry, the hair along a goat's spine stands on end. Its body hair sometimes stands up too.

Goats tend to move in family units with a herd, with an alpha female (herd queen) leading the way and the herd's alpha buck (herd king) bringing up the rear.

When alarmed or angered, goats fluff up their coats and a ridge of hackles stands up along their spines.

GOAT PHYSIOLOGY

Temperature: 101.5–103.5°F (38.6–39.7°C) for adult goats
Heart rate: 70–90 beats per minute
Respiration rate: 12–20 breaths per minute
Ruminal movements: 1–1.5 per minute
Life span: Usually 10–12 years (the known record is 23)

DOES

Age at puberty: 5–10 months
Breeding weight: 60–75 percent of adult weight
Heat cycle: Every 18–23 days
Heat duration: 12–36 hours
Ovulation: 12–36 hours after onset of standing heat
Length of gestation: 144–157 days
Number of young: 1–5 (twins are the norm)
Breeding season (seasonal breeders): August–January
Breeding season (aseasonal breeders): year-round

BUCKS

Age at puberty: 5–8 months
Primary rut: August–January (but most meat breeds will breed
year-round)
Breeding ratio: One adult buck to (up to) 30 does

Goats pant when the temperature and humidity rise. Humidity, more than temperature, stresses goats; when it's humid, milk production plummets, goats lose weight, and interherd hostilities increase.

Goats spend more time ranging and feeding when it's neither hot (above 90 degrees) nor cold (below 50 degrees). They drink more water as the temperature rises and considerably less in colder weather.

While goats require clean drinking water (they'll usually refuse any other kind), they lose very little body moisture through panting and elimination. Except for lactating does, goats can temporarily get by on comparatively little water if the need arises.

Goats hate rain, water puddles, and mud, but given sufficient feed, water, and shelter, they easily withstand most weather extremes.

What This Means to You

The closer you can duplicate your goats' natural preferences, the more likely you are to keep them content. Contentment translates into healthy goats that won't constantly challenge your fences.

Social Order

A well-defined pecking order exists within every goat herd, large or small. Where a goat stands in this hierarchy depends on its age, sex, personality, aggressiveness toward other goats, and the size of its horns (or lack thereof). Unweaned kids assume their dam's place in the order and often rank immediately below her after they've been weaned.

Newcomers battle to establish a place in the herd. Fighting for social position is conducted one-on-one; established herd members don't gang up on a newcomer. When fighting, males shove one another, butt, and side-rake opponents with their horns; does usually butt, jostle, and push. To butt each other, combatants stand a few feet apart, facing one another. They rear up and swoop forward, down, and to the side to smash their horns against those of their opponent. Bucks do not back up and charge in the manner of rams.

Did You Know?

- Goats have a 320- to 340-degree panoramic field of vision and 20- to 60-degree binocular vision.
- In a German study conducted in 1980, male goats were tested to determine their capacity for color vision. They were able to distinguish yellow, orange, blue, violet, and green from gray shades of similar brightness.
- Another study proved goats can distinguish among bitter, salty, sweet, and sour flavors. They have a high tolerance for bitter tastes, hence their preference for bitter-tasting twigs and plants such as wild herbs and weeds over tender, tasty, cultivated pasture grasses.
- Goats also have a keen sense of smell. Does recognize their newborn young by their smell, and buck scent is an aphrodisiac for does.

Other forms of aggression include staring, horn threats (chin down, horns jutting forward), pressing horns or forehead against another goat, rearing without actually butting, and ramming an opponent's rear end or side. There is very little in-fighting in a static herd, however, where each member knows and accepts its place.

The herd is led by its queen, an old alpha doe who has head-butted, shoved, and threatened her way to the top of the heap. She's rarely (if ever) challenged and she remains queen until she's removed from

Angry goats rear before swooping toward each other and bashing foreheads.

the herd or she becomes too feeble to lead, at which time the position is often assumed by one of the old herd queen's daughters. When the herd moves, it's because she leads them. When she stops, they stop. When she eats, the rest eat, and when possible, they choose whatever the herd queen is eating.

Most of the year, any bucks in the group, including the herd king or alpha male, defer to the herd queen. When breeding time rolls around, however, the herd king assumes leadership. The king breeds all of the does — underling bucks are simply out of luck. Yet bold ones constantly challenge the herd king; a herd queen generally outlasts many youthful kings.

Even young kids (especially bucklings) make horn threats at one another.

What This Means to You

Because the queen leads the herd to food, if you're the person who dishes out their eats, your goats will soon consider you the ultimate herd queen. Then your goats (even the real herd queen) will rarely challenge you and you can lead them wherever you want them to go.

If you insist on driving goats from the rear, however, they'll perceive you as a two-legged herd king—not a good position to be in when there are male challengers in the group.

A GOAT'S CROWNING GLORY

- According to a series of studies reported in *Hierarchy in a Group of Goats,* by the Polish biologist Marcin Tadeusz Gorecki, for does, bucks, and even wethers, horns are the most important factor in determining a goat's place in the herd's hierarchy. Size and age are also important, but color is not.
- Although most goats have horns unless they're disbudded, some goats are polled, meaning they're naturally hornless. Breeding one polled goat to another is not without risk; a certain number of kids from such breedings will be intersexed: they'll have sexual apparatus that is neither fully male nor fully female. Some goat producers treasure intersexed goats (also called hermaphrodites) as teasers — infertile males that can be run with does and, by their sexual activities, signify when does are ready to be bred.
- Goat horns are filled with blood vessels. If you nip off more than the very tip, a horn will bleed profusely.
- 4-H and FFA project goats generally cannot have horns; therefore, kids produced for this market should be disbudded at an early age.
- It's unwise to handle goats by their horns; a snapped-off horn is a red-flag injury. When goats must be handled by their horns, grasp horns near their base, where they're strongest, and don't exert more pressure than is absolutely necessary.
- As young goats' horns emerge and begin to grow, the outer surface sometimes seems to peel. Not to worry; this is normal. As the horns grow longer, they'll smooth out again.

Breeding Behavior: Bucks

Although most bucks reach puberty at about five to ten months of age, bucklings as young as eight weeks old have been known to impregnate their mothers.

Both seasonal and aseasonal breeding bucks enter rut as autumn approaches and stay in rut through the first months of winter. The volume and motility of a buck's semen is greatest at this time, even among aseasonal breeders.

Bucks that live peaceably with other bucks most of the year become highly intolerant toward one another during rut. Each considers himself herd king, so if bucks from different herds somehow get together, there will be bloodshed. Bull-stout fences, preferably with wide aisles between them, are usually necessary to keep apart two breeding bucks.

During rut, scent glands located near a buck's horns secrete strong-scented musk (when bucks rub their foreheads on a person or object, they're spreading their scent), and bucks become more vocal. They spray thin streams of urine along their bellies, on their front legs and chest, and into their mouths and beards. Bucks also twist themselves and grasp their penises in their mouths. They sometimes masturbate on their bellies and front legs and then sniff themselves and flehmen (see box on page 82).

A buck puts on a display when courting a doe in heat. He'll enurinate (spray urine on himself), paw the earth, and emit an amazing series of vocalizations known as blubbering. If the doe doesn't flee, he'll assume a slight crouch and sashay closer with his head slightly extended and his tail curled up across his back. He'll gobble, flick and wag his tongue, and, if his lady love

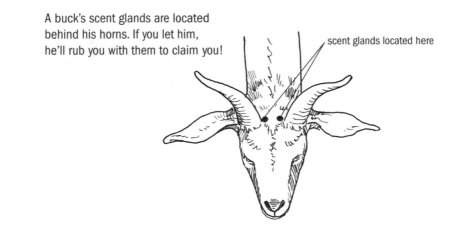

A buck's scent glands are located behind his horns. If you let him, he'll rub you with them to claim you!

scent glands located here

If a doe in heat allows it, a courting buck will move quite close to her, with his head extended and his tail curled across his back.

allows it, he'll sniff and nuzzle her sides and her vulva. If she urinates, he'll catch urine in his mouth and flehmen.

When covering a doe, the average buck makes several false starts before ejaculating. When he ejaculates, he thrusts his hips forward and then leaps off the ground while twisting his head back and to one side. Small, young, or otherwise inexperienced bucks sometimes fall off the doe onto the ground.

When he's finished, the buck will lick his penis and flehmen, rest a short while, and start again.

THE FLEHMEN RESPONSE

The flehmen response, also called the flehmen position, flehmen reaction, and flehmening, is the grimace most ungulates, cats, and a few other mammals make to examine scent.

When he flehmens (which he does to determine if a doe is in heat), a buck flips his upper lip back to expose and draw odorants into his Jacobsen's organ, a pheromone-detecting organ located in the roof of the mouth. All sexes flehmen after sniffing unusual scents, especially the urine of other goats.

PLEASE DON'T CALL THEM BILLIES!

In an effort to counteract the ornery, tin can–eating billy goat image of yore, most meat goat producers, dairy goat enthusiasts, and goat hobbyists call their goats bucks and does (in the manner of deer) instead of the old-fashioned terms billies and nannies. If you're tempted to do otherwise, remember this joke:

> **Q**: What's the difference between a buck and a billy?
> **A**: $500 to $1,000.

What This Means to You

Because tame bucks frequently view humans as herd members, women working around bucks in rut should be aware of and discourage courtship behavior. The buck who perches his forefeet on his female caretaker's chest and gobbles in her face means no harm, but being bowled over by a 200-pound hunk of amorous, smelly buck is nothing to joke about. Some otherwise easygoing bucks, once in rut, can also perceive human male caretakers as competition and react accordingly.

No matter his (or her) size, age, or level of experience, a human handler should at all times remain aware of bucks in rut.

Breeding and Kidding Behavior: Does

From early autumn through midwinter, seasonal breeders (such as dairy breed does) cycle, or come in heat, every 18 to 23 days; aseasonal breeders (most meat breed does) cycle all year round. Ovulation occurs 12 to 36 hours after the onset of standing heat.

Does are stimulated by the appearance and scent of a buck. Given their choice, most does choose an old buck with impressive horns over a younger rival. They rub their necks and bodies against him to court his favor. A fully receptive doe stands with her head slightly lowered, her legs braced, and her tail to one side. She may urinate when he sniffs and nuzzles.

Signs of Heat

- Interest in nearby bucks
- Increased activity level (especially fence walking)
- Loud and insistent vocalizing

- Tail wagging (sometimes referred to as flagging)
- Increased urination
- Mounting other does and being mounted by them
- Decreased appetite
- Lower milk production

If impregnated, a doe enters anestrous and most does stop coming into heat (though a small percentage of does allow themselves to be bred throughout their pregnancies).

One hundred and forty-four to 157 days after the onset of her last standing heat, a pregnant doe gives birth to from one to five offspring. First-time moms usually birth a single kid; mature does of meat breeds usually produce twins or better.

A day or so before kidding, the average doe becomes fretful and anxious. A few hours before giving birth, does living in herd situations usually leave the group and seek a secluded birthing place. They may or may not be accompanied by a friend (usually a doe's grown daughter or her dam).

Signs of Impending Kidding

- **A filled udder.** It may be stretched so tight that the organ appears shiny and teats may jut out to the sides (this is called a *strutted udder*).
- **Mucus.** A glob of mucus may cling to the doe's vulva or hang from it in strings; with some does, stringing begins a few days prior to kidding.
- **Restlessness.** An hour before kidding, most does begin circling and pawing the ground or bedding in an attempt to hollow out a soft spot on which to give birth (this is referred to as *nesting behavior*). They lie down, get up to reposition themselves, flop down, rise, and repeat this ritual many times.
- **Introspection.** Once nesting begins, many does' attention turns inward. They murmur softly to their unborn kids in a vocalization used only prior to kidding and during the first few days of the new kids' lives.

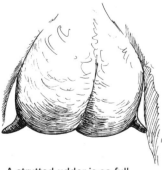

A strutted udder is so full that it seems ready to pop.

THE BUCK RAG

According to some veteran goat breeders, it's possible for doe owners to tell when she is in heat by letting her sniff a buck rag.

To prepare a buck rag, wipe a terry cloth washcloth all over a buck in rut, concentrating on the scent glands near his horns and on any areas stained with urine. When it's permeated with stink, quickly pop the washcloth in a tightly sealed glass jar and keep it there until you think your doe is in heat. When you do, open the jar in her presence, pull out the washcloth, and let her sniff it. Theoretically, when she smells that yummy aroma, she'll "show" as if the buck were actually there.

Kidding Behavior

As hard labor begins, the doe rolls onto her side to push. She may struggle to her feet, turn, and reposition herself; this helps put her kids into correct birthing positions.

After each birth the doe rises and licks clean her newest arrival. Does initially recognize their kids by scent; cleaning is an important element in the bonding process and it also physically stimulates the newborn kids.

In a best-case scenario, the doe's afterbirth passes within a short time after her final kid is delivered. Does frequently eat the afterbirth. Some goat producers allow this; however, because they represent a choking hazard, others remove the membranes as soon as they're delivered.

What This Means to You

By recognizing signs of heat, producers who house bucks separately from does (and it's a good idea; that way, exact breeding dates are known and kidding dates can be calculated with accuracy) know precisely when to bring a doe to a buck.

Further, a goat producer who recognizes signs of impending birth can be on hand when the event occurs. This is important; we'll talk more about attending kiddings in chapter 13.

Behaviors of Newborn Kids

Most kids struggle to their shaky legs within 10 to 30 minutes after birth and begin actively seeking a teat. Hardy-breed newborns such as Kiko and Kiko-cross kids are faster to rise and feed than kids of other breeds. This is a plus, because kids should ideally ingest colostrum (first milk) within the first hour or two after their birth.

Neonatal kids are wired to seek darkness (in places such as the armpits and groin) and warm, bare skin. Weak or disoriented kids can be helped

SEPARATING THE SHEEP FROM THE GOATS

Sheep and goats have a great deal in common. For instance, both contract many of the same diseases and parasites. Because of these similarities, sheep producers can often relatively easily switch to raising meat goats. There are definite differences between the two species, however. Here are a few worth mentioning.

- Because their chromosome counts are dissimilar, sheep bred to goats or vice versa (and when sheep and goats are kept together, this sometimes happens) rarely produce viable offspring.
- In most breeds, sheep's horns sweep back, then curl down and out into tight outward spirals next to the animals' faces, while goats' horns generally sweep back and then out or up and then out (although Boer goats' horns spiral somewhat too).
- Breeding together two polled goats often produces intersexed offspring. This doesn't occur when breeding polled sheep to other polled sheep.
- Goats are often bearded; sheep never grow beards.
- Goats' tails flip up over their backs; sheep's tails hang down and are generally carpeted with wool.
- A pastured sheep's diet consists of 90 percent grass, forbs (weeds), and herbs and only 10 percent browse. A pastured goat prefers a ratio of 40 percent grass to 60 percent browse.
- Goats are a lying-out species; sheep are not.
- Bucks reek during rut; rams have no odor.

by holding them near their dam's udder, although most resist a teat placed directly into their mouths.

Kids kneel to nurse and they butt their mothers' udder to facilitate milk letdown. A rapidly wagging tail means a kid is suckling milk. After feeding, contented kids take a nap. A kid that constantly calls out, suckles, or probes at its dam's udder isn't finding enough to eat. Without intervention, such a kid could die.

- Most sheep flock closely, making it easy to work them with herding dogs. Goats flock too, but when harried, they try to break and run, making it difficult to herd goats with dogs.
- When goats fight, they rear and swoop down toward each other to butt heads; when rams fight, they back up a distance before lowering their heads to charge.
- Goats are more likely to climb on things and scale fences; they're more lithe and agile than sheep.
- A ewe's more complex cervix makes breeding via vaginal artificial insemination nearly impossible, while does are routinely bred this way.
- Goats are more likely to have triplets, quads, and quints (although a few sheep breeds such as the Polypay, Finn, and Romanov are geared to have litters too).
- Goats require additional copper in their diets, while copper is toxic to sheep in relatively small amounts.
- Most goats are less parasite-resistant than are sheep.

RAM   BUCK

A kid kneels to nurse. To facilitate the flow of the dam's milk, it butts her udder.

Goats are a lying-out species, like cattle and deer. Does place their kids in what they deem safe spots, then go off to graze, returning roughly four to six times a day to feed their hidden young. When kids begin nibbling at pasture and browse, they start following their mothers; this can occur in a day or two or up to a week from the time they're born. Scientists conducting sheep and goat cross-fostering studies noted that lambs raised by does grew faster than kids raised by ewes or even kids raised by their own dams because the lambs trailed their surrogate mothers all day and thereby suckled more often. Conversely, ewes were constantly flustered by their foster offspring when the goat kids refused to shadow their woolly moms. While kids do follow their mothers, they are far more independent than lambs, going off in bunches to play right from the start.

When kids — even older, weaned ones — are handled for unpleasant procedures such as banding or when they're given injections, they shriek a high-pitched distress call sure to bring out close neighbors in droves. Some Boer kids sound like human children being tortured. You'll be flabbergasted the first time you hear this bloodcurdling shriek.

What This Means to You

By understanding neonate behavior, producers are better able to judge when neonatal kids require assistance to survive. They're also less likely to panic when pasture-kept does are spotted out feeding without their newborn kids in tow.

Goat Handling 101

Even healthy goats that never leave home must be handled for routine chores such as hoof trimming, vaccinating, and deworming. A simple handling system makes these tasks safer for handlers and less stressful for goats.

In our small-scale operation, we use a simple chute constructed of a set of parallel stock panels with gates at each end, through which our goats enter and exit their nighttime housing every day. Because they aren't afraid of this arrangement, it's easy to shut the gates and trap goats three at a time for deworming or vaccinating, and it's simple to halter the one whose hooves you need to trim while he's trapped with his herd mates in the chute.

Some small-scale goat ranchers may prefer a catch pen to a chute. A catch pen is a small, well-fenced area into which several goats can be herded (or lured with feed) to make it simpler to catch the goat you want. To catch a goat

In a large-scale operation, goats are sorted by funneling them through a chute.

In a small-scale operation, the easy way to catch a flighty goat is to corner her in a pen made from fencing.

in a pen, extend your arms to create a visual barrier. Approach slowly and calmly. Quietly ease it into a corner. Once it's cornered, if the goat isn't wearing a collar, move quickly to capture it by the base of its horns or place one hand under the goat's jaw and lift up on its head while cradling its rump below the tail with your other hand. To move forward, let the goat's head down a little and lift up on its tail, but be prepared for the goat to move explosively.

Large-scale producers who prefer to do things quickly and easily may prefer more complex handling systems incorporating crowding pens (to funnel goats into a chute), chutes, and sorting pens. These homemade or commercially constructed systems can be built of wood, stock panels, metal, or pipe. See chapter 8 for more about handling systems.

Flight Zones and Blind Zones

It's important to note that goats have long memories and rarely forget a negative experience. Because stress equates with illness in goats, no matter what sort of handling facilities you use, it behooves you to work calmly and quietly, taking care not to frighten or manhandle goats.

All animals have a flight zone; we humans call this "personal space." If you get inside a goat's flight zone and the goat isn't comfortable having you there, it's

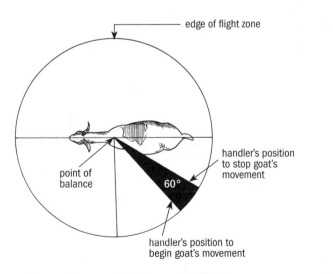

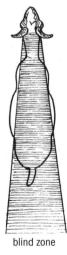

If you enter a goat's flight zone behind his shoulder, he will move forward. Adjust your position as shown to stop movement.

A goat's blind zone isn't very large because of his wide peripheral vision.

going to move. The size of an individual animal's flight zone varies according to how tame it is as well as the size of the enclosure you're working in. Flight zones decrease dramatically in cramped quarters. To move a goat but prevent it from bolting, work just inside the animal's flight zone, but no closer.

All animals also have a point of balance; for goats, it's located at the nearest shoulder. Enter its flight zone at a point behind its shoulder and a goat will move forward; enter at a point ahead of its shoulder and the goat will back up or turn around and move the other way.

All animals have a blind zone behind their shoulders. Because goats have wide peripheral vision, their blind zone isn't nearly as wide as ours is. Goats always want to know what's behind them, however; they don't like to have unknowns in their blind zone. If you invade its blind zone, a goat is probably going to move forward. Calm goats move forward at a walk; frightened goats are likely to scatter. Keep things low-key and relaxed to save yourself exasperation and wasted time.

Goats are notoriously hard to drive, but they are very, very good at following a leader. If they accept you as herd queen, get out front and let them follow you. If you're not the herd queen, a pail of grain or other goodies will instantly make you queen for a day.

Always remember that goats are startled by loud noises and sudden movements, and they prefer to move forward out of darkness into light. Poor depth perception makes them wary of shadows and of abrupt light-into-darkness situations. They also prefer to move from confined places into more open ones, uphill rather than down, and into the wind rather than downwind. They don't care to cross water or slippery surfaces and dislike passing through narrow openings.

It's easier to lead goats with a bucket of grain than to drive them.

Goat Safety

When pressured, bucks sometimes turn and fight. Don't think you can hold your own with an enraged 200-pound buck. If you try, you're very likely to get hurt.

If you're pressured by a buck, watch him but don't stare into his eyes and don't stomp your feet; these are perceived as challenges.

Avoid entering pens with bucks in rut. If you must be in a pen with a rutting buck, be prepared. Keep your eye on him at all times and don't allow him to position himself between you and your only escape route. If you know he's bad-natured, carry a small shock prod with you; it sounds cruel, but if you need it, it could save your life. A spray bottle filled with one part chlorine bleach to ten parts water can be used instead of a prod; a well-aimed spritz in his face will usually deter a buck without injuring him. If threatened, hold your hands out to your sides and make yourself look as large as you can.

In a pinch, grab a feisty buck by his beard, hang on tight, and walk him to the fence, where you can release him as you slip outside. The buck-by-the-beard ploy works just about every time, so don't forget this procedure.

Watch Those Horns

It's easy to get hurt by horns, even when your goat doesn't mean to hurt you. When working closely with a horned goat, perhaps to train it or doctor its face, make a small hole in each of two old tennis balls and push one onto the

ZONING OUT

Some goats slip into a catatonic state when they're deeply frightened. This response in goats was first recorded by Ivan Pavlov, the Russian behaviorist who conducted classical conditioning experiments using dogs. The right thing to do is back off and allow the goat time to recover on its own. When it has returned to normal, proceed with whatever you were doing but in a quieter, more goat-friendly manner.

When confined in close quarters such as a holding pen, goats become aggressive toward one another. Don't confine small or timid goats with large, aggressive individuals; horned goats with hornless ones; or a bunch of bucks in an area just large enough for them to battle.

tip of each horn. When you're finished, pull them off and save them to use another day.

And be careful of those horns if you straddle a goat to restrain it; Boers, especially, can do a great deal of harm to you if they back up when you least suspect it.

And Keep an Eye on the Kids

Your human kids, that is! Small children are naturally attracted to barnyard activity, but they need to be supervised when playing around goats. A frightened goat can easily bowl over a toddler; a doe with newborns that wouldn't think of butting you might consider your six-year-old fair game; and the buck's pen should be considered strictly off-limits to those who are neither old enough nor strong enough to protect themselves from aggression or amorous advances.

Tennis balls can prevent injury to a handler during close work.

Producers Profile

Al and Kirsten Kosinski, Black Bell Acres
Alton, Missouri

Al and Kirsten Kosinski raise registered Savannas and Savanna-Boer market kids at Black Bell Acres, a 67-acre farm nestled in the foothills of the Ozark Mountains. Savanna goats are a small but important part of a diversified farming venture launched when they left their 19-acre Michigan farm in search of fairer weather and enough land to grow and market grass-fed meat, fresh goat milk, and free range eggs. In addition to meat goats, the Kosinskis raise Highland cattle and Alpine dairy goats, Katahdin-Dorper cross hair sheep, and heritage chickens, ducks, and geese. When we talked to Kirsten about their diversified farming venture, this is what she said.

Q *How did you become involved with meat goats in general and Savannas in particular?*

A We were looking into different breeds and as beginners, we wanted something easy to care for, hardy, and parasite resistant. We joined a local goat club and heard about Savanna goats, so we went to visit a breeder in Arkansas and fell in love with them.

Q *Are you satisfied with the breed?*

A Yes. We have two registered does and one buck and have never had to worm the buck and one of the does. We don't feed them grain and they do well on pasture.

Q *What segment of the meat goat industry do you serve?*

A We also cross our Savanna buck with our Boer and Boer cross does for percentage breeding stock and market kids. Our goats are handled and kept in a stress-free environment, and lots of sunshine, fresh clean water, and plenty of grass create lean, tender, and flavorful meat. When we started, we sold USDA-inspected meat at the West Plains farmers market but now we sell goats on the hoof.

We are also building our herd of registered Savanna goats. Our goal is to produce high quality Savannas, not just quantity. We are a small operation and this allows us time to nurture the animals as they grow and mature. We strive to improve the genetics of our goats and produce a hardy and healthy Savanna goat for anybody to enjoy.

Q *Is the goat market strong in your part of the country? Do you think there's room for more meat goat producers?*

A Goat meat seems to be in high demand. The United States is still importing goat meat, so I think yes, there is room for more goat producers.

Q *Do you have any advice for someone interested in starting with meat goats?*

A Go to a breeder with a good reputation. We do that with all of the breeds we raise. For instance, we went to Oklahoma to buy our Highland cattle and we bought our goats from Carl Langle in Viola, Arkansas, who owns the oldest herd of purebred Savanna goats in the country.

7

Goats on the Go

AN ENTIRE CHAPTER ABOUT TRANSPORTING GOATS may seem like overkill, but it's not. Travel-stressed goats die of enterotoxemia, pneumonia, dehydration, and a host of other stress-related maladies. Pregnant does can abort from traveling stress and lactating goats can contract mastitis. Immature doelings hauled with bucks can get bred and kids can be trampled. But none of this has to happen; whether you're bringing home your first goats, traveling to shows for the hundredth time, or trucking kids to market, it's important that you transport goats with utmost care.

Stress and the Traveling Goat

Goats don't handle changes very well. Add the noise, confusion, and crowding that are part and parcel of trucking and you're going to have some sorely stressed goats.

In *Lowering Stress in Transported Goats,* Craig Richardson, animal care specialist writing for the Ontario Ministry of Agriculture and Food, says that transportation stress factors are of two types: short-acting factors that cause emotional stress and long-acting factors that cause physical effects and tend to accumulate over the duration of a trip.

Short-acting factors stressful to hauled goats include unfamiliar surroundings, unfamiliar traveling companions, and unstable footing. Long-acting, often cumulative factors include noise, vibration, being thrown against the sides of the vehicle or other animals, fatigue from standing, insufficient food and water, and extremes of temperature and humidity.

Plan Ahead

To cut down on in-transit injuries, post-transport weight loss, and stress-induced illness, plan ahead to make hauling easier on your goats. Here are some strategies to consider.

- **Avoid hauling sick or injured goats and late-gestation does.** Start out with sound, healthy animals and do your best to keep them that way. And because goats are social creatures (a solitary goat is a stressed-out goat), haul along a companion if you possibly can.
- **Map the route in advance.** According to Richardson, braking and cornering cause 75 percent of all in-transit falls; crossing bumps and acceleration cause the rest. Stop-and-start driving causes hormones and blood components to fluctuate and can send heart rates up to twice their norm. If the most direct route means dealing with rush-hour traffic or possibly hitting red lights at scores of stoplights, it's always wiser to choose a longer but easier route.
- **Factor rest stops into a long journeys.** Locate points along the route where you can safely offload goats at least once every 24 hours. Offer them familiar feed, water them, and check each goat for signs of injury and excessive stress. Pack a well-appointed first-aid kit (see box on page 98), and to cut back on digestion-related illness, bring along enough probiotic gel to dose each goat at least once a day.
- **Load goats with care.** Don't drag them along by choke collars or lift them by their horns, head, ears, hair, or legs. Cover the nonslip floor of their conveyance with an adequate amount of dust- and mold-free bedding or improvise by covering a slippery floor with several inches of damp sand and topping that with conventional bedding. If you're loading goats at night, provide plenty of interior lighting; goats move more easily from darkness into light than the other way around.
- **Allow sufficient room for each goat.** On short jaunts, each animal should be allotted enough room to stand without constantly slamming into another one or the sides of the vehicle; on longer hauls, goats need enough space to lie down comfortably.
- **Prefit the conveyance with interior dividers.** Bucks and other aggressive goats act out their traveling annoyance by butting and horn-hooking timid and smaller goats. Interim dividers enable you to partition goats into compatible groupings based on sex, size, age, and/or aggressiveness. Barring that, halter and tie bucks and other unruly goats, making sure tie

ropes are long enough to allow them to get back up if they fall but not so long that they can choke or get hopelessly tangled.

- **Cut down on vibration.** Whatever sort of conveyance you choose, use plenty of bedding, and reduce noise levels by padding gates, loading chutes, and partitions with pieces of rubber matting or old blankets.
- **Allow enough time to drive carefully.** Accelerate slowly and smoothly and do your best to stop that way too. Ease up on the gas well ahead of turns and don't take corners too abruptly. Factor in load checks, also; stop 20 minutes after departure to check your load and at least once an hour after that.

Weather Kills

Weather extremes head the list of long-term stressors. The upper heat tolerance for goats is 95 to 104 degrees. As the temperature and humidity rise inside their traveling compartment, goats become restless and start to pant. Goats standing with their necks extended and who are gasping for air are in deep distress, as are goats who have fallen and refuse to get back up.

Heat Exhaustion and Loss of Body Heat in Goats

Your most important response to a goat suffering from heat exhaustion is to cool him — fast. The best way is to hose him down using plenty of cold water and then place the goat in front of a fan to finish cooling. Just getting the goat wet isn't enough; you need to use *cold* water and then let it evaporate.

When traveling in hot, steamy weather, pack along a cooler containing jugs of ice-cold water and bagged ice cubes. Gently trickle cold water on the back of the head of distressed goats. Wrap ice cubes in cloth and hold them in the goats' armpits or groins.

If a great deal of summer travel is on your agenda, spring for a power inverter that connects to your truck battery and allows a full-size, household box fan to operate on truck power. You'll be glad you have it when one of your valuable goats collapses en route to a show.

In a pinch, cool down a distressed goat as quickly as you can, then stuff him in the cab of your truck with the air conditioner on high and drive around so the air conditioner works more efficiently. It isn't easy if he's a big one, and it definitely turns the heads of passersby, but it's better than losing the goat.

To compensate for sizzling, steamy weather, reduce load capacity by 15 to 20 percent (overcrowding rapidly leads to excessive heat buildup); create additional ventilation by opening windows or replacing solid upper walls of your conveyance with sturdy, closely spaced pipe or heavy wire mesh; travel only at night or during cooler morning hours; and keep the number and length of stops to the barest minimum. Never, under any circumstances, park a truck or trailer loaded with goats in direct summer sunlight. Offer water to parched goats whenever and wherever you stop.

BUILD A TRAVELING FIRST-AID KIT

Put together a traveling first-aid kit to augment the kit you keep at home in the barn. Pack it in a lidded five-gallon plastic pail (restaurants and fast-food establishments will often give you one for free) and keep it in your truck or trailer at all times. If you use any supplies from the kit, replace them as soon as you get home. Having a well-equipped first-aid kit and knowing how to use its contents can make the difference between life and death when you're on the road and far from the closest vet. At the bare minimum, a kit should include:

- Betadine or a similar microbicide to flush fresh wounds
- Sterile gauze sponges
- Telfa nonstick absorbent pads to cover wounds
- Several individually wrapped sanitary napkins to use as pressure pads to stop heavy bleeding
- Several rolls of Vetwrap or a comparable self-adhesive bandage
- A roll of sterile gauze bandage, 2½ inches wide
- Rolls of adhesive tape, 1 and 2 inches wide
- Antibiotic ointment
- Saline solution and topical eye ointment
- A rectal thermometer and lubricant
- Tweezers or a hemostat
- Probios or a comparable probiotic gel
- Banamine — a prescription injectable pain reliever and anti-inflammatory drug acquired from your vet — or children's aspirin for pain
- Some Band-Aids and nonprescription pain relievers for yourself

Cold kills too. Goats (especially kids) are susceptible to frostbite and loss of body heat. It's important to keep in-transit goats dry. If the weather is cold or damp, cover openings in your conveyance to protect goats from rain and wind chill; add more bedding; allow extra space so goats can move away from chilling wind. Fit your goats with snug goat blankets to keep their bodies warm.

Choosing a Hauling Conveyance

Goats can be safely hauled in goat totes (boxes or cages designed to be set in the bed of a pickup truck) and truck cappers, stock and horse trailers, and even in crates in vans and pickup trucks. When choosing your animals' conveyance, keep these points in mind:

- Goats *can't* be safely hauled in completely enclosed cargo trailers or any sort of conveyance with openings large enough for a goat to escape (and a determined goat can squeeze through amazingly small openings).
- Goats can die quickly from carbon monoxide poisoning, so make absolutely certain engine exhaust can't enter the area occupied by goats.
- Goats are very susceptible to respiratory problems and can easily expire from overheating. Good ventilation is an absolutely essential element of goat conveyance design.
- Poor footing leads to scrambling, physical injuries, and rampant stress. Damp wood and some rubber flooring are as slick as glass once they get wet. Invest in nonslip trailer mats; they quickly pay for themselves many times over. Use dust-free disposable bedding on top of the mats (dusty bedding contributes to respiratory ills).

The Goat Hauler's Secret Weapon: Dog Crates

One item every goat producer needs is an extra-large dog crate. Heavy-duty wire dog crates are best; they're ideal for in-home housing for bottle babies (goats are social creatures and they like to see what's going on) and they're great for hauling young goats in the back of a van. Choose one with a flip-open top and a swing-open end door that can be tightly secured. Pad the crate with old, washable bed blankets. It's that easy!

Fastidious folks with nicer vans prefer airline-style plastic crates that don't allow urine and worse to leak out onto the floor. Wire crates need not have leaking problems, however; the trick is changing bedding before it gets nasty — and having a place (such as a nylon duffel bag on a rooftop luggage carrier) to keep soiled blankets until they can be laundered.

On the plus side, crates fit nicely in even the smallest vans if the backseat is temporarily removed. (You may want to put down a plastic tarp for leakage insurance.) They're also useful for hauling goats in the bed of a truck. While plastic airline-style crates are best for this use, you can use a wire crate and cut down on wind stress by wrapping a blanket around all but the end facing the tailgate of your truck. Substitute a plastic tarp if it's raining, but avoid plastic coverings when it's sizzling hot. Don't haul goats in crates in freezing weather; a single goat in such a close enclosure can't generate enough heat to stay warm.

Goat Totes

A number of leading livestock trailer and goat-handling-equipment companies make slide-in truck bed carriers designed especially for goats. Some are enclosed units with adjustable ventilation features such as securely screened, sliding windows; others are essentially cages with sliding back doors.

Enclosed units are heavier, sometimes requiring a fair amount of muscle power to hoist them onto the truck; look for aluminum slide-ins if weight is an issue. Enclosed units are more secure than goat cages and provide more protection when hauling valuable goats.

Goat totes are the safe and popular way to transport a small number of goats.

Cage-type slide-ins work well too, but should be used with covers to protect hauled goats from inclement weather. And double-latch those doors! It never hurts to secure them with additional clips; a goat cage door that opens when you're traveling 60 miles an hour is a recipe for disaster.

Working with a Livestock Transporter

In many cases, transporting your own goats isn't cost effective. If you're relocating to a new home 1,000 miles away and taking your goats with you, if you've purchased your dream doe and her kids via a distant Internet auction, or if you buy your first passel of goats from a breeder in another state, you may elect to pay a livestock carrier to transport your animals.

Most professional and weekend transporters (goat owners headed to shows or transporting goats of their own who are willing to carry along a few extras for pay) post their trips on goat-related listservs and e-mail lists (another great reason to subscribe to these). You can also locate professionals by searching for livestock transport with your favorite Internet search engine.

Transporters usually charge a hefty fee for the first animal picked up per stop and considerably less if you ship more than one. Obtain estimates from at least three companies; prices vary dramatically from hauler to hauler.

Unless you contract for the whole load, routes are rarely direct and your goats may be on board for days — or occasionally weeks. Because goats are easily stressed, it's important to choose a transporter who knows goats and goes the extra mile to keep them safe and well. When comparing rides, ask these questions:

- What sort of trailer and hauling unit does the hauler use? What are his or her contingencies in case of breakdown?
- Is he or she willing to separate your goats from other livestock with a partition? If so, will your animals have any sort of physical contact with other goats and sheep en route or when off-loaded at rest stops?
- Will the hauler pack along a supply of your goats' accustomed feed (and feed it to them)?
- What is the hauler's policy for dealing with sick or injured goats?
- If something goes amiss, will the hauler contact both shipper and receiver so both are informed? Will the hauler phone if he or she is running late? Can you and other involved parties reach the hauler by cell phone en route?

When shopping for livestock transporters, ask for references and take time to check them. Better still, post to listservs and ask other subscribers to contact you (off-list) with their recommendations or tales of woe.

Find out what sorts of tests and health papers are necessary for your goats' interstate shipment and have the paperwork in hand when the hauler arrives to pick them up.

And don't necessarily choose the cheapest transporter or the one who does the most advertising; find one you're comfortable with based on experience or other goat owners' recommendations. Transporting goats is fraught with risk; it's important to find and hire the best livestock hauler you can.

8

Meat Goat Housing & Facilities

YOU COULD BUILD A BIG FANCY ENCLOSED BARN to house your goats, but meat goats are healthier (and happier) in simpler surroundings. In fact, if you already have buildings on your ranch or farm, you can probably convert them to house goats. Kept in dry, well-ventilated, draft-free surroundings, goats are unusually adaptable creatures.

Shelter

Perhaps the best (and certainly the simplest) solution is to build or buy enough inexpensive, low-slung, three-sided field shelters to house your herd. If you put the shelters on skids, so much the better — you can move them to greener pastures as you move the goats. The best field shelters for goats are 5 to 6 feet tall in front, sloping to 3 to 4 feet high in back, and 8 to 10 feet deep. Build them as long as you like, allowing 12 to 15 square feet of bedded sleeping space per adult goat. The open side should face away from prevailing winter winds (a southwest exposure fully utilizes the winter sun as a source of heat). Such a shelter's low headroom holds in heat and its shallow depth allows for good ventilation. You'll have to duck your head when you check your goats, and these shelters are bothersome to clean with a pitchfork, but from a goat's-eye view, they're sheer perfection.

In hot climates, build just a roof and framework and enclose three sides with welded-wire cattle panels. Open-air shelters provide shade and rain protection in the summertime; when winter comes, you can enclose the fenced sides with sturdy plastic tarps. It doesn't look elegant, but it works!

Think outside the box when choosing goat shelters. Goat producers house their charges in some innovative structures. Refurbished chicken coops, calf hutches, southern-style carports with semi-enclosed sides, ready-made portable livestock shelters such as Port-a-Huts and PolyDomes, Quonset huts, and cattle-panel and plastic-tarp hoop houses can be used to house meat goats. Or you could keep them in a regular barn.

If you loose-house your goats in a barn, allow 20 square feet per adult indoors and 25 square feet per goat in an adjoining exercise yard. Make certain that ventilation is adequate and that goats aren't caught in chilly drafts.

Packed dirt or stone floors are easier on goats' legs and feet than concrete; slatted wood floors provide great drainage, but they tend to rot. Whatever floors are made of, top them with four to six inches of absorbent, dust-free bedding (sand, straw, poor-quality hay, shavings, rice or peanut hulls, even shredded paper) and clean it as needed through the summer months. When winter comes, follow a deep bedding system by removing surface wetness and badly soiled areas, then adding more bedding atop the remainder. Decomposing manure pack provides a source of winter heat and makes a fine addition to your compost heap come spring.

Goats kept in poorly ventilated, tightly enclosed winter housing tend to suffer from respiratory ailments such as bronchitis and pneumonia. Don't shut all the doors and windows — meat goats (even thin-skinned Boers) can handle considerable subzero cold. Heating goat housing is rarely an issue. If goats get sick or kids are born during a cold snap, blanket them instead of heating the barn. If you do use a heat lamp, make certain it's securely tied high enough (don't hang it by its cord) to prevent bedding from igniting from the heat or curious goats from pulling it down.

Goats needn't live in a palace to be happy. Simple field shelters make ideal housing.

Pens

In addition to fenced pastures, you'll need some pens. Pens can range in size from a length of cattle panel on each of four sides to an acre or more and are used for kidding and weaning; to hold special goats such as pets, old duffers, 4-H wethers, and bucks; for new goats fresh from quarantine that haven't yet been turned out with the main herd; and for thin goats that need more feed to gain some weight. The list goes on — no matter how many pens you build, you'll wish you had more.

A PLAYHOUSE FOR THE KIDS

In addition to full-size shelters, consider putting out mini-shelters just for kids. These give youngsters a safe place to escape from aggressive adult goats and to nap or hang out with one another.

- Buy your kids an extra-large, two-piece, easy-to-clean plastic doghouse (readily available at pet shops, some farm stores, or discount stores). One house will accommodate a passel of newborn kids or two or three 10- to 12-week-old youngsters.
- Better yet, get them a Dogloo, an igloo-shaped doghouse by Petmate. Dogloos have separate floors but their walls and roofs are molded in a single piece, making the units practically indestructible. Kids love to climb and bounce off these structures, and they come in an array of sizes; one even has an offset entrance to keep cold winds at bay.
- Use heavy tin snips to cut little doorways in overturned, large plastic cattle mineral tubs. These can be perfect shelters — just be sure to set a rock atop each one; they tend to take flight in a strong wind.

Pens are usually built using stout woven wire. When they are small or price is no object, welded-wire cattle panels work better still. Cattle panels are the fence of choice for buck pens.

Cattle Panels to the Rescue!

One thing you'll realize when you bring home goats is that cattle panels are the goat farmer's best friend. Sometimes called stock panels, cattle panels are pre-fabricated lengths of sturdy mesh fence welded from galvanized quarter-inch

A BLANKET FOR THE GOAT

A number of companies sell handsome winter blankets designed for goats; however, blankets designed for horse foals work every bit as well.

- When measuring a goat for a foal blanket, hold a cloth dressmaker's tape at the center of his brisket and measure straight along his side to directly below his tail.
- It's better to err on the small side than to buy a blanket that's way too big.
- Choose a high-denier blanket. (Denier is a unit of measure for the density of fabrics.) A 1,000-denier blanket is considerably heavier, much more snag-proof, and infinitely more durable than one made of 300-denier fabric. You should especially spring for higher denier if you have horned goats.
- Opt for a blanket with stretchy elastic hind leg straps; if you buy one that doesn't have them, add them yourself (design your own or buy them ready-made from a horse supply outfit). Most goats don't like blankets and they'll try to shed them; leg straps help stabilize a blanket and keep it on the goat.

steel rods. Most cattle panels are 52 inches tall and built using 8-inch stays; horizontal wires are set closer together near the bottom of the panel to prevent small livestock from escaping. Cattle panels are usually sold in 16-foot lengths, which can be trimmed to size using heavy bolt cutters.

Sheep panels are like cattle panels but are manufactured in 34- and 40-inch heights, and their horizontal wires are set even closer together. Both cattle and sheep panels are ideal for fencing goats out of the garden and for fabricating extra-stout pens and corrals.

Kids are even easier to "blanket" than adult goats. There are many inexpensive kid coats to choose from:

- Visit a Goodwill store or Salvation Army or shop garage sales to find inexpensive secondhand sweatshirts and cardigan sweaters for human babies. Depending on the size of the kid and the clothing, you may want to trim the sleeves from garments before dressing the goat. Wool cardigans are perfect kid warmers in cold-winter climates; simply feed the kid's forelegs through the armholes and button the sweater up its back.
- Dog sweaters make cute, inexpensive kid coats. Don't hesitate to buy a pullover style because it doesn't look as though it'll fit — those sweaters stretch much more than you'd think. Choose wool for subfreezing weather, but in warmer winter climates and for spring or fall kiddings, standard acrylic dog sweaters work well.
- Cut neck and leg holes in a man's extra-large winter wool sock to fashion a snug-fitting, warm kid coat for a preemie.
- Buy kid coats from Hoeggers or Caprine Supply or watch for them listed on eBay. Some ready-made goat coats are cute as can be, and they really keep the little ones warm.

Cattle-panel fences are the strong, safe, but rather expensive way to enclose goats.

Utility panels are the toughest of all. They're fabricated using 4-inch by 4-inch spacing and are welded from extra-heavy-duty 4- or 6-gauge rods in a full 20-foot length. They come in 4- to 6-foot heights and are ideal to use for buck pens.

Panels designed specifically for goat and sheep are available in 36-inch and 40-inch heights and in 4- to 6-foot lengths — just right for building V-type hay bunks and round-bale hay feeders for small ruminants. These or cut-to-size standard cattle panels are ideal for constructing easy-to-lift walk-through gates; kidding jugs (small pens where a doe and her kids are kept for 24 to 72 hours after birth); and small, temporary pens.

One drawback to standard cattle, sheep, and utility panels is that the raw end of each rod is very sharp. To make these panels more user-friendly, smooth each rod end with a rasp to remove its razor edge. Premier 1 sheep and goat panels (see Resources) are smoothed at the factory.

Feeders

You can buy ready-made goat troughs and feeders or make them yourself to fit your own needs. Be aware that all good goat-feeding apparatus has one thing in common: Each is designed to discourage goats from wasting feed. Being the fastidious creatures they are, goats won't touch hay, grain, or minerals they've peed or pooped in; conversely, they certainly don't mind peeing and pooping in their feed. Kids complicate the matter by "nesting" in accessible hayracks and grain or mineral feeders — and they don't vacate their nests when nature calls. It's the producer's mission to prevent these unsanitary and wasteful practices. When considering goat feeders, remember these points:

You definitely need them. Feeding grain or hay off the ground contributes to parasitism and disease, not to mention excessive waste.

Feed all the goats in a herd at the same time; otherwise timid goats won't be allowed to eat. Allow 16 inches of feeder space per adult horned goat and 12 inches if your goats are disbudded or polled.

Choose grain feeders you can move by yourself. Several smaller feeders are more manageable than a single extra-bulky one, and rubber and plastic feeders are lighter than metal models. During wet spells, a mud wallow will form around your feeders if you don't move them every few weeks; this and manure buildup in feeding areas contributes to parasitism and disease.

For small groups of goats, consider feeders you can hang on a fence and remove after the goats have eaten. A good inexpensive homemade one is crafted of a length of eight-inch PVC pipe cut in half lengthwise and attached to the fence with S-hooks.

V-shaped feeders like this one prevent goats from soiling and wasting expensive hay.

You can mount grain feeders six inches higher than your tallest goats' tails and provide booster blocks or rails for their front feet to stand on.

If you have many goats and opt for a large-capacity feeder that operates on the gravity-feed principle, choose one specifically designed for goats. (The hoppers in cattle models are often spacious enough for kid goats to use as bunks.) Choose an easy-to-clean model. Because you won't be moving it often, to keep muck and mire from forming around the feeder, set it on a concrete pad extending at least six feet out from the trough.

For feeding small rectangular bales of hay, V-shaped hay feeders with welded-wire sides or vertical or diagonal slats work better than models with horizontal slats or bars.

Make an effective, inexpensive fence-line hay feeder by wiring the bottom of trimmed-to-size 4 × 4-inch welded-mesh wire panels to an existing fence and adding sturdy wire spreader arms at the top.

When feeding large round bales of hay, save a huge amount of waste by limiting goats' access with a feeder ring designed for goats.

Loose mineral feeders should be placed where they won't be rained on, either in a building or under a canopy. Because most mineral mixes are 10 to 25 percent salt and salt is highly corrosive, plastic mineral feeders are the durable choice.

Build an effective gravity-feed, loose mineral feeder by gluing a Y-type PVC cleanout plug to the bottom of a 3- or 4-foot length of 4-inch PVC pipe,

Homemade PVC mineral feeders work as well or better than any commercial feeder.

with the arms of the Y facing up. Permanently cap the stem of the Y and attach a removable cap to the top of the tube. Add minerals through the top and hang it at chin height on your smallest goats.

Watering Devices

Most goats consume between ½ and 3 gallons of water a day, depending on weather conditions (goats require more water during the sizzling summer months), mind-set, and whether or not they're does in either the latter stages of pregnancy or in milk. Lactating does have the highest requirements — and to prevent urinary calculi (mineral-based stones that can block male goats' urinary tracts and lead to untimely death), it's important for wethers and bucks to drink a great deal too.

Yet when water supplies are contaminated with goat droppings, algae, dead birds, or bugs, leaves, and other debris, goats drink only enough fluid to eke by. If the water you serve your goats isn't appetizing enough for you to drink it, you can bet the goats will shun it too. Empty scummy or contaminated tubs, tanks, and troughs and scrub or spray their inner surfaces with a solution of one part chlorine bleach to ten parts water. If you're in the market for something to hold water for your goats, choose a series of small troughs or automatic waterers over one or two mega-model tanks; the littler ones are easier to clean.

Place water-filled tubs and buckets in the shade during the summer months. This helps inhibit algal growth, and because the water stays fresher,

WATERING TUBS FOR PENNIES

If you keep cattle, you probably have empty plastic mineral lick tubs sitting around. Smaller sizes are shallow enough to make first-rate kid watering troughs and the taller ones are handy for watering adults.

If you don't have cattle, talk to someone who does. Or ask a friendly clerk at your favorite feed store; he may know someone who has a pile of tubs to give away.

the goats will drink more. When temperatures soar into the 90s or higher, freeze ice in plastic milk bottles and submerge one in each trough or tub; your goats will appreciate this treat. Refreeze them overnight and they'll be ready to use again by mid-morning.

During the winter, keep water supplies from freezing by installing bucket or stock tank heaters — but remember to encase the cords in PVC pipe or a garden hose split and taped back together with duct tape; if you don't, goats may gnaw through the cord and electrocute themselves.

When kids are present, water depth should be no more than 14 inches lest a frolicking youngster leap or fall in and drown. *Never* use five-gallon recycled plastic food service buckets or other narrow, deep containers in kidding pens.

Handling Facilities

As we learned in Goat Handling 101 in chapter 6 (pages 89–90), ranchers with fewer than 30 goats can usually manage with a catch pen and possibly a simple handling chute, but larger producers may prefer a full-scale goat-handling system.

Large-scale goat operations often need a complete goat-handling system.

FENCES FOR HANDLING FACILITIES

When you work in close proximity with wild or reluctant goats, you don't want them to leap out and head for the hills, so make handling-facility fences tall and stout. If your goats are tame, however, cattle panel fences work just as well at a fraction of the cost (just remember to use a rasp to smooth their sharp wire ends to prevent injury to you or your animals).

Ready-made systems are crafted of sturdy steel and are designed to be moved, if you want, but homemade systems work well too. Here are some plans to consider.

Build Your Own Goat-Handling System

While there a few plans for goat-handling systems on the Internet, systems (and other equipment) designed for sheep work well for meat goats too. The Canada Plan Service (see Resources) offers an array of free, downloadable sheep equipment PDFs, including two excellent handling system plans.

- One offers a sorting and treatment corral suitable for a herd of 700 to 1,000 does, more or less. The fencing is arranged to handle goats quickly, quietly, efficiently, and with a minimum of stress to both goats and operators.
- Another is a working chute and one component of a corral system for sorting, weighing, spraying, and other essential management operations. Fitted into the corral, it extends from a crowding pen so the herd can be gently funneled through the chute. Sliding gates at the entrance and exit ends of the working chute allow for traffic control, while a pair of sorting gates hinged to the sides of the chute provide for rapid, three-way sorting.
- Most university Extension services can furnish easily adaptable sheep building and equipment plans upon request, including plans for constructing working corrals; cutting gates; chutes; and complete, portable handling systems. Some of these plans can be found online.

Fencing

Toss a pail of water at your fence;
if the water goes through, so will your goats.

— Wise old goat keeper's saying

Good fences make good neighbors.

— The wise old goat keeper's next-door neighbor

Keeping goats fenced is a giant headache. If there is a weakness in the fence line — even just one — intelligent, inquisitive goats will find it and go over, under, or through the fence and be on their merry way. Furthermore, if goats can get out, predators can get in and that can be catastrophic. Plan on spending a chunk of cash to build goat-resistant fences; they're the costliest part of getting into goats.

Several types of fences work well with goats; the most commonly used materials are woven wire, barbed wire, electric wire, and portable fencing such as IntelliTape and ElectroNet.

Woven Wire

Woven wire, also called wire mesh or field fence, is the fence of choice for goat farms. It's made of horizontal lines of smooth wire held apart by vertical wires called *stays*. The spacing of its horizontal wires is usually closer near the bottom of the fence. The vertical stays in standard woven wire are 6 inches apart; horned goats tend to get their heads stuck in this type of close-spaced fencing, so *goat net* with 12-inch stays is a better choice. Its major failings: Its price and the effort it takes to install it.

Correctly installed, woven wire is the most secure affordable goat fencing, making it ideal for perimeter or boundary fence. Four-foot-high woven wire will contain most goats. Installing one or two strands of barbed or electric wire above woven wire helps keep predators out of your herd's enclosure.

One drawback to woven wire is that goats lean into it and then stroll along from post to post, scrubbing their sides. While this is a great way for goats to shed winter hair or get rid of a vexing fly, it's very, very hard on the fence. To prevent this, most producers install a strand of offset hot electric wire on the inside of woven-wire fences at goat shoulder height. It works!

High-tensile woven wire costs more than standard woven-wire fencing but it's rust-resistant, it sags less, and it's lighter in weight.

Nowadays, most woven wire comes with galvanized (zinc) or aluminum coating. Both styles are further classified as class 1, 2, or 3 wire; the higher the number, the thicker the coating and the more durable the fence. Class 1 galvanized woven wire will show signs of rusting in 8 to 10 years; class 3 fencing begins to rust in 15 to 20 years. Aluminum-coated wire resists corrosion three to five times longer than galvanized wire with the same thickness of coating. Because a major part of fencing cost is its installation, it's best to buy the longest-lasting wire you can afford.

Woven wire is sold in 20-rod rolls (a rod is 16.5 feet; 20 rods is 330 feet) and is usually supported by wood or steel posts erected at 14- to 16-foot intervals. Wood posts come in treated and untreated varieties. Treated posts will last 20 to 30 years; untreated ones at least 2 years, depending on the type of wood they're made from.

The strength of wood posts increases as its top diameter grows larger: a 4-inch post is twice as strong as a 3-inch post; a 5-inch post is twice as strong as a 4-inch post; and so on. Corner and gate posts should have a top diameter of at least 8 inches; brace posts should be 5 inches or more in diameter; and

Woven-wire fencing is standard in the goat industry, but it's important that all openings be small enough to prevent horned goats from putting their heads through them or large enough to allow them to remove their heads without getting stuck.

HEAD-IN-FENCE SYNDROME

One of the most annoying (and scary) things about raising horned goats is that they love to get their heads stuck in woven-wire and welded-wire cattle panel fences.

Goats, from new kids to the oldest codgers in the herd, subscribe to the "grass is greener on the other side of the fence" school of thought. When they force their heads through woven-wire or cattle panel fencing to reach for that irresistible morsel, their horns act like fishhooks, and sometimes they can't back out.

A trapped goat is at the mercy of its herd mates (who may beat the stuffing out of the head-stuck goat), predators (that tasty head is stuck where it's readily accessible — not a pretty picture, but it sometimes happens), the elements (rain, snow — goats get their heads stuck no matter what the weather's doing), and themselves (stuck long enough, goats become shocky, their systems shut down, and they die).

What to do for this pathetic creature?

- Switch it to a pen or pasture surrounded by goat fence (12-inch stays allow enough wiggle room for goats to extract themselves from this mess), electric, or barbed wire.
- Duct-tape a length of PVC or wooden dowel to a goat's horns so it can't put its head through the fence. It looks silly, but it works. Give the creature a month or so, then remove the apparatus and see if it's reformed. Some goats learn, some don't.
- If it's one of the latter group, *sell her.*

line posts, can be any diameter 2½ inches or more, but the bigger the posts the stronger and more durable the fence.

Steel posts come in U-bar, studded Y, punched channel, and studded T types, but they're all commonly referred to as T-posts. Although they lack the eye appeal of wood posts, T-posts are fireproof, long-wearing, lighter in weight, and relatively easy to drive. They also ground the fence against lightning when the earth is wet. They do tend to bend if larger livestock leans against them or you back into one with your tractor. Expect unbent T-posts to last 25 to 30 years.

Harvest Your Own Fence Posts

If you live where Osage orange, eastern red cedar, or black locust trees grow and you have more time and energy than money, you can cut your own wooden posts. Untreated red cedar and black locust posts last 15 to 25 years and untreated Osage orange posts last, at minimum, 25 years. Other suitable do-it-yourself fence-post woods are untreated white oak, ironwood, honey locust, catalpa, mulberry, and hickory. Untreated, each of these lasts at least 15 years.

Whichever species you choose, you must peel each post before setting it in the ground. It's easiest to peel them in the spring, when rising sap causes bark to loosen.

Barbed Wire

Barbed wire is the classic stockman's fence. Properly installed, it can hold goats. It's easier to work with than woven wire and it's less expensive, but drawbacks are that it's not as predator-proof as other goat fences and, for safety reasons, it must never be electrified to give it more teeth. (An animal could get tangled in the barbs and charger pulses might kill it before you discover it's in trouble.)

THE TAG TELLS ALL

When buying woven wire, read the tag; the numbers printed on it will tell you how it's made. For instance, 10-47-6-9 fencing has 10 horizontal wires; it's 47 inches tall; there's a 6-inch spacing between stay wires; and the fence is made of 9-gauge wire.

8 in.

8 in.

6 in.

5 in.

5 in.

4 in.

4 in.

4 in.

3 in.

Goats can be contained by barbed-wire fences, but it's important to use enough strands to keep them in.

Barbed wire consists of two or more strands of regular or high-tensile smooth wire twisted together with two or four sharp barbs added at 5- to 6-inch intervals. It's sold in 80-rod rolls in a variety of sizes and patterns. It can be erected on wood or steel T-posts set 12 to 20 feet apart. It's sometimes reinforced by installing twisted wire or plastic stays between each set of fence posts.

It takes eight to ten strands of 15½-gauge or better barbed-wire fence to contain goats, with the lower strands set closer together than the ones near the top. A working configuration for a nine-strand barbed-wire fence: place the bottom strand three inches above the ground, each of the next three

THE SKINNY ON FENCING

One of the best guides to buying materials and installing fences is the information-packed Premier1 fencing catalog (see Resources). Because Premier1 handles all styles and sizes of livestock fencing and chargers, its catalog/guide to choosing and installing fencing is remarkably unbiased. You'll find helpful fencing articles in each of its archived newsletters, too.

strands four inches above the one below it, the fifth wire five inches above the strand below it, the sixth wire five inches above the fifth, the seventh wire six inches above the sixth, and each of the top two strands eight inches above the one below it.

Additional strands of barbed wire can be used to beef up existing fences and render them suitable for goats. Despite its attractive features, however, barbed wire can cause serious injuries to the livestock it contains and should never be used to fence areas sometimes inhabited by flighty animals such as horses.

Electric Wire

Electric fencing wire is marketed in aluminum, regular steel, and high-tensile steel varieties. For permanent fencing, high tensile is best.

High-tensile, smooth fencing wire is installed using wood or T-posts and plastic insulators. It comes in 11- to 14-gauge models and has a breaking strength of roughly 1,800 pounds. High-tensile fencing is extremely durable, it's relatively easy to install, and it can be stretched extremely tight without breaking. Strong corner and end braces are needed to install it, as well as the tensioners and strainers used to keep it bowstring-taut.

Where high-tensile fences aren't needed, standard steel or aluminum wire will do the trick. Aluminum wire is the better of the two; it's rustproof and so easy to work that you can shape it with your bare hands, and it conducts electricity much better than does steel wire.

Goat fence failures happen when the fencers (also called chargers or energizers) that power them aren't up to the job. Fencers are sold by their voltage (4,000 volts and up will hold most goats) and the number of joules they put out (a joule is the amount of energy released with each pulse). One joule will power 6 miles of single-wire fence; 4½-joule fencer will energize 20 to 60 acres, depending on the length of the fence and the number of wires that are used in its construction. Replace old-style "weed burner" fencers with modern, low-impedance models that neither short out when damp vegetation touches a wire nor spark grass fires during times of drought.

Install five to seven wires when building electrified goat fences. A five-strand fence is adequate for most trained goats. To make one, install the first four wires nine inches equidistant from one another, starting nine inches from the ground, and the fifth wire a foot above the fourth.

Electric fences must be properly grounded or they'll lose their punch, yet according to Susan Schoenian (in "Facilities and Equipment for Commercial Meat Goats," see Resources) an estimated 80 percent of the electric fences in

Once trained to respect it, most goats can be kept inside electric fencing, but it's important to use an energizer with plenty of bite and to keep it operational at all times.

the United States are improperly grounded. A minimum of three 6- to 8-foot ground rods should be used with each fence charger, and it's important to follow the charger manufacturer's instructions when installing them.

Tips for Building Better Electric Fences

- **Buy an adequate fencer.** The box will tell you how many miles of fence it charges, but that's the greatest length of one strand of fencing operating under tip-top conditions. Think big. The more powerful the fencer, the fewer problems you'll have, so pick a model that packs a punch.
- **Read the instructions.** Mount solar fencers directly facing the sun; install ground rods as directed in the user's manual. Don't wing it. Unless you install a fencer correctly, it can't do the job it was designed to do.
- **Use quality insulators.** Sunlight degrades plastic, so choose high-quality insulators, preferably a brand treated to resist damage caused by ultraviolet (UV) light.

THE WALKING RULER

Forget the metal tape measure the next time you build a fence. When installing barbed-wire or electric fencing, wear old jeans and use a felt-tipped marker to make fencing-height marks on your pant legs; it will save you a great deal of time.

- **Don't skimp on wire.** The larger the wire, the more electricity it can carry.
- **Don't space wires too closely.** To get the most from your energizer, keep wires at least five to seven inches apart, even near the ground, even when building goat fences.
- **Never electrify barbed wire.** A goat could get tangled in the barbs and continuing pulses from a big charger might kill the animal before you discover it's in trouble.
- **Train your goats to respect electric fences.** Electric fences are psychological rather than physical barriers; in order for goats to respect them, the animals need to get mightily zapped. Place untrained goats in a fairly small area equipped with a hefty fencer, and then entice them from the sidelines with a pail of tasty grain. Better, save metal tops from canned goods, poke holes in them, switch off the fencer and use wire to attach the lids to the fence at goat-nose height, smear both sides with a layer of peanut butter, then add the goats. Once most goats have been shocked, they tend to avoid electrified fences like the plague.
- **Buy a voltmeter and use it.** A good one costing in the neighborhood of $50 to $75 will help you keep your fences hot. Check the voltage every day. If the fence runs low on voltage or shorts out, you'll want to know it and correct the problem right away.

Portable Fencing

Polywire, polytape, and rope-style electric fencing used with step-in poly or fiberglass posts makes fine interior fences and can be moved around with

Polytape-style electric fencing is both attractive and effective, but always twist the tape several times between each post to reduce wind resistance.

Portable net fencing is ideal if you plan on moving it frequently — but it must not be used to contain horned goats or small kids!

ease. Wide, flat tapes offer high visibility — especially important when your fences must hold horses too — but flat tapes whip around in the wind more than rope-style temporary fencing. To minimize whip, twist the tape once or twice between fence posts rather than installing it flat.

Net-style portable fencing is wildly popular with grass-based farmers and goat keepers interested in controlled grazing. There are two basic types: those with built-in posts and the kind without. Most roll up onto easy-to-use reels, making moving these lightweight fences a breeze. Net fencing has one serious drawback, however: unless they're trained to respect electric fences before encountering soft net fencing, horned goats can sometimes entangle their horns in the mesh, get repeatedly shocked, stress out, and even die. The same thing can happen to vulnerable young kids who get their heads tangled in the fencing.

EIGHT USES FOR PORTABLE NET FENCING

- To subdivide pastures for rotational grazing
- To erect small pastures near the barn for special-needs animals such as bucks or does with new kids
- To create limited grazing for animals in quarantine
- To keep out coyotes and marauding dogs (a few brands also work well against foxes)
- To keep in livestock guardian dogs
- To fence steep, rocky, or otherwise uneven land
- To use as boundary fences on rented land
- To make lanes for moving goats without assistance

9

Feeding Meat Goats

PLEASE NOTE THAT WHILE I'VE BASED the following material on extensive research conducted at universities and on experiences and advice shared by longtime producers of quality meat goats, other sources you consult may not agree with this information. Goat nutrition is a tremendously complex subject, so my ultimate advice is to discuss feeding protocols with your county Extension agent (he's familiar with the needs of goats raised in your region, as well as the feed available to them), a goat nutritionist, or someone in your locale who is successfully producing the sort of goats you'd like to raise.

The Caprine KIN (Keep It Natural) Diet and Your Goats

A load of misinformation surrounds the feeding of goats. The average non-goat owner believes goats eat anything — but they don't. Furthermore, some of the things we feed goats aren't good for them, and opinion varies widely regarding what they *should* be fed. At some point, as a goat producer, you'll have to choose which feeding philosophy best suits your needs and those of your goats. This is mine.

The nutrient requirements of goats are affected by an animal's breed, age, weight, production status (breeding buck, lactating doe, packing wether), and a host of other variables such as climate and the mineral content of the land where the animal grazes or its feed is grown. It's difficult for a layperson to calculate a proper diet based on so many variables, so the best advice is to leave it up to the experts. One possible diet is based on high-quality forage (browse, pasture, or hay); a mineral supplement formulated for the type of forage you feed and the area in which you live; and, when conditions dictate, a judicious amount of properly balanced, commercially bagged grain.

Here is my rationale for supporting such a diet: The ruminant stomach is superbly designed to efficiently digest the cellulose (also called fiber, a major structural carbohydrate in plants) in forage. It doesn't, however, adapt well to starch-rich concentrate (grain) diets, which predispose goats to nutritional diseases as diverse as bloat, ruminal acidosis (see the box on acidosis, page 129), enterotoxemia, goat polio, listeriosis, laminitis, and urinary calculi. Many classes of goats, among them dry does, bucks, and wethers, thrive on a diet of quality forage. Mineral supplements can be used to balance the ration and protein blocks to boost protein intake, all without exposing goats to the dangers of concentrate overconsumption.

When grain is needed (and it's a must for growing youngsters, late gestation and lactating does, and hardworking bucks of certain breeds), it should be nutritionally balanced and fed in moderation. While some producers have the knowledge, time, and inclination to carefully balance the nutrients in home-mixed feed (Langston University's online interactive Total Mixed Ration Calculator can help do-it-yourselfers balance them correctly), most of us don't. Bagged goat feeds may initially seem more expensive than home mixes, but because they're backed by nutritional research, feeding bagged feed according to the producing company's directions minimizes incidences of growth-stunting nutritional deficiencies and potentially catastrophic metabolic diseases such as pregnancy toxemia, urinary calculi, and enterotoxemia. In the long run, feeding commercially formulated, bagged feed makes very good sense.

The Goat's Digestive System

To understand why forage should comprise the major part of the caprine diet, you need to understand the ruminant stomach.

All true ruminants (goats, sheep, cows, deer, bison, giraffes, and the like) have stomachs composed of four chambers: rumen, reticulum, omasum, and abomasum.

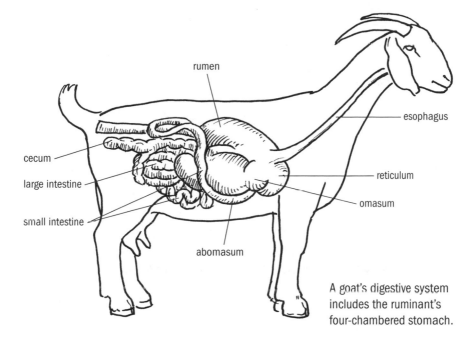

A goat's digestive system includes the ruminant's four-chambered stomach.

The Rumen

The *rumen* is the first and largest of the four chambers. It's essentially a roomy fermentation vat. As newly consumed feed mixed with saliva enters the rumen, it separates into layers of solid and liquid material. Later, when the goat is resting, it regurgitates a bolus of food (its *cud*) and rechews the material more slowly, then swallows it again. During second (and subsequent) chewings, feed is mixed with more and more saliva, which, being alkaline, helps regulate rumen pH at a healthy 7.0 to 7.8. (See the box on page 127 for more information on pH.)

The Reticulum

Once sufficiently liquefied, macerated feed exits the rumen through an overflow connection called the *rumino-reticular fold* and enters the *reticulum*, where fermentation continues (methane is continuously produced by bacteria and protozoa in the rumen and reticulum; this is the source of the stinky belches goats emit), until it passes through a short tunnel into the omasum.

KIDS DON'T FUNCTION LIKE ADULTS

- A newborn ruminant's rumen is small and undeveloped. A kid functions as a single-stomached *(monogastric)* animal until it has nibbled enough of its environment (dirt, mama's hair, even manure; kids explore everything with their mouths) to introduce the symbiotic bacteria and protozoa its rumen needs to begin functioning.

- A nursing kid's head-back, chin-up posture causes a band of tissue called the esophageal groove to close, thus creating a direct tubular connection from the esophagus to the abomasum. Therefore, it's important to feed bottle kids in the normal nursing position with their heads back and their noses in the air. Otherwise, milk enters the nonfunctional rumen — where, some researchers believe, it stagnates and contributes to floppy kid syndrome (see chapter 10, page 149).

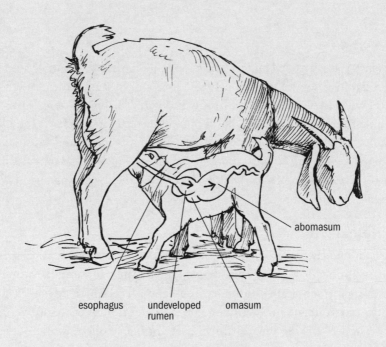

esophagus   undeveloped   omasum   abomasum
rumen

WHAT IS pH?

The pH scale ranges from 0 to 14.0. It measures the acidity or alkalinity of a solution. A pH of 7.0 designates a neutral solution; a pH of less than 7.0 means the solution is acidic; a pH of more than 7.0 means the solution is alkaline. Typical pH readings are (from acid to alkali):

Battery acid: to 0.5
Human gastric acid: 1.5 to 2.0
Lemon juice: 2.4
Vinegar: 2.9
Coffee: 5.0
Milk: 6.5
Pure water: 7.0 (neutral)
The contents of a healthy goat's rumen: 7.0 to 7.8
Seawater: 8.0
Hand soap: 9.0 to 10.0
Lye: 13.5

The Omasum

The *omasum* (also called the *manyplies*) is divided by long folds of tissue that resemble the pages of a book; they are sometimes referred to as leaves. These leaves are covered with small, fingerlike structures called *papillae* that help decrease the size of feed particles, remove excess fluid, and absorb any volatile fatty acids that weren't absorbed in the rumen.

The Abomasum

The fourth chamber, the *abomasum*, is considered the goat's "true" stomach because it's where digestion occurs much as it does in the human stomach. The abomasum, like the omasum, contains a large number of folds that increase its surface area many times over. The walls of the abomasum secrete digestive enzymes and hydrochloric acid, which quickly lowers the pH of feed slurry from about 6.0 to around 2.5. Protein is partially broken down in the abomasum before digestive slurry is shunted along to the small intestine.

The Small and Large Intestines

As semidigested feed enters the small intestine, it's mixed with secretions from the liver and pancreas that push the pH back up from 2.5 to between 7.0 and 8.0, making it possible for the enzymes in the small intestine to reduce any remaining proteins into amino acids, starch into glucose, and complex fats into fatty acids. As this occurs, muscular contractions continue to push small amounts of material through the system and into the large intestine, where resident bacteria finish digesting the mix.

What This Means to You

A goat's rumen contains billions of bacteria, protozoa, and other microbes that feed on carbohydrates in the goat's diet and convert them into volatile fatty acids (the goat's primary source of energy); without them, the animal would die.

   One type of microorganism is configured to digest cell-wall carbohydrates (cellulose, hemicellulose, pectin, and lignin; e.g., fiber). This functions best in a neutral environment (about 7.0 pH). Another digests cell-soluble carbohydrates (sugars and starches; e.g., concentrates such as grains and commercially bagged feeds); these prefer a more acidic environment (5.0 to 6.0 pH).

Did You Know?

- A goat makes an estimated 40,000 to 60,000 jaw movements per day as it chews and rechews regurgitated food.
- A goat's four-compartment stomach occupies three-quarters of its abdominal cavity.
- An adult goat's rumen can hold three to six *gallons* of feed.
- Microorganisms in the rumen and reticulum generate methane. Methane causes the incredibly stinky odor in goat burps.
- A newborn kid's abomasum (the fourth and "true" stomach) comprises 50 to 70 percent of its total stomach area; at eight weeks, only 30 percent; and by adulthood the abomasum comprises only 9 percent of the goat's stomach.
- The intestinal tract of an adult goat is about 100 feet long and can hold three gallons of partially digested food.
- Up to 50 percent of each goat dropping may be of microbial origin.

What the goat consumes affects the pH of its rumen. When it eats fibrous forage, it cuds more and releases more saliva. Goat saliva contains bicarbonate, and this buffers the rumen, thus preventing it from becoming too acidic. Due to the low fiber content in concentrates, however, less chewing is required, so less saliva is produced. Also, the starch in grain is rapidly fermented, resulting in skyrocketing ruminal acid production. When a goat eats too much grain, the rumen's pH plummets and forage-digesting microorganisms die, clearing the way for lactic acid–producing microbes to proliferate and produce more acid. The result: Acidosis (see box below).

In short, this fine-tuned machine called the goat's digestive system requires long-stem fiber to operate in the pink. A forage-based diet is the basis of a healthy, productive goat herd. Forage doesn't make up the entire picture, however. Some goats, especially growing kids and late-gestation or lactating does, require grain supplementation due to their increased energy and protein needs. This is where bagged concentrates shine. Feed them according to their manufacturers' directions and don't add feed elements that will disrupt a product's carefully calculated nutritional balance.

ACIDOSIS

Acidosis, also called lactic acidosis or grain overload, occurs when rumen pH falls below about 5.5. As pH continues to decline, rumen microbes die and ruminal action decreases or ceases altogether. Symptoms include depression, dehydration, bloat, racing pulse and respiration, staggering, coma, and then death. Due to permanent damage to the rumen lining and intestines, survivors generally fail to thrive.

Treatment for acidosis: Call your vet — don't delay!

Prevention: Goats should be fed with an emphasis on forage rather than grain. Make all feed changes gradually, allowing rumen microbes time to adapt to a new diet. Lactating does consuming high levels of concentrates can be provided with free-choice baking soda to nibble as the need arises (don't feed baking soda to bucks; it's said to contribute to the formation of urinary calculi).

Goats and Poisonous Plants

No matter where you live in the United States, there are doubtless poisonous plants growing in areas where your goats feed. This may or may not be a problem, depending on whether or not:

- **Your animals eat them.** Poisonous plants aren't necessarily attractive to goats. Pasture-wise goats seem to know intuitively which plants they can safely consume. And because many poisonous plants taste nasty — acrid or extremely bitter — goats won't eat them unless they're hungry and it's either eat them or starve.
- **They eat these plants in dangerous volume.** Many "poisonous" plants are simply toxic. Unless they're eaten in massive quantities or over a length of time, they do no damage. Because goats are nibblers, they often don't consume enough of poisonous plants to cause themselves harm.
- **They consume the poisonous part of the plant.** In many cases, only a portion of a plant is poisonous — its roots or wilted leaves or seeds — or the plant is poisonous only at certain stages of its growth and goats don't eat it at that time of the year.
- **They're immune to the compounds in a given plant.** Some poisons are species-specific. Pigs thrive on acorns, sheep consume larkspur, and humans eat milkweed shoots, yet all of these plants can poison a goat.

According to various sources, at least some parts of the following plants are poisonous (or toxic) to goats at some point in the plants' annual growth cycle:

aconite	bloodroot	death camas
amaryllis	blue cohosh	dogbane
arrowgrass	boxwood	elderberry
avocado	broomcorn	false hellebore
azalea	buttercup	hemp
bagpod	celandine	horse nettle
baneberry	cherry	hydrangea
barberry	chokecherry	Indian hemp
bellyache bush	cocklebur	Indian poke
bittersweet	coffeeweed	inkberry
black locust	cowbane	jimsonweed
black snakeroot	crow poison	johnsongrass
black walnut	crowfoot	Klamath weed

lantana	philodendron	snapwort
larkspur	poison hemlock	St.-John's-wort
laurel	poke	stagger grass
lily of the valley	potato	staggerbush
lobelia	rattlebox	thornapple
lupine	rattleweed	tomato
maleberry	rhododendron	velvetgrass
marijuana	rhubarb	white cohosh
milkweed	rock poppy	wild black cherry
monkshood	senecio	wild parsnip
nightshade	sesbania	wolfbane
oak	sevenbark	yellow jasmine
oleander	snakeberry	yew

A Forage-Based Diet

Forage can mean browse, grass, hay, or a combination of all three. The best is arguably browse. Given room to roam, goats spend their days tasting here, nibbling there, harvesting material from cellulose-rich plant species chock-full of the nutrients goats need, and that's how it ought to be.

Goats lacking access to a plentiful supply of pasture or browse should be fed free-choice, quality hay composed of grasses and/or legumes that have been air- and sun-dried until only about 18 percent of their moisture remains. When selecting hay for your goats, remember the following.

- **Quality is more important than type.** Properly baled coastal Bermud grass a is higher in protein and other digestible nutrients than rained-on, first-cutting alfalfa. Meat goats require 10 to 16 percent crude protein and 60 to 65 percent total digestible nutrients (TDN) in their total diet, which consists of all the hay, browse, pasture, grain, and supplements they consume. Many types of hay fulfill these needs, especially when fed with a reasonable amount of grain or with goat-specific protein blocks.
- **Hay should smell fresh, never sour or musty.** There should be no sign of dust or mold. Animals that breathe mold spores from hay can develop permanent lung damage.
- **Reject new bales that seem unusually heavy for their size or that feel warm.** The problem is that these weren't fully dry when baled. Dampness

generates heat, which in turn causes the hay to mold or, worse, triggers spontaneous combustion.

■ **Ask the seller to open several bales so you can evaluate the hay inside.** Most types of hay bales should be green inside. Hay bales that are yellow or brown inside are composed of hay that's been rained on or sun-bleached in the field before baling. In either case, the nutritional quality will be lower than that of properly put-up hay; however, don't worry about slight discoloration on the outside of bales, especially if they've been stacked in the sun.

■ **Buy as much hay at one time** as you can properly store to minimize metabolic disease caused by changes in hay type or quality.

■ **Buy tested hay or test the hay you bale or buy to determine its crude protein and TDN.** Use a hay probe; you can buy one at a farm store or borrow one from your county extension agent (he or she will give you a list of forage-testing laboratories and a test kit too). Sample three or four bales representative of all your hay: shove the probe into the center of each bale to extract each sample, fill a plastic bag with about a pound of combined sample material, complete the paperwork, and mail the sample to a certified laboratory. Depending on the mailing method you use, expect the results back in two to seven days. The basic test should run $15 to $25.

■ **Establish terms before you finalize the purchase:** prices, delivery terms, who is responsible for hay that doesn't meet your standards, and so forth.

■ **Don't depend on word of mouth or the classifieds when buying hay.** For a wider choice of sellers, pick up a list of hay suppliers at your county extension office or peruse hay lists online at the USDA Farm Service Agency's HayNet website (see Resources).

Variables Affecting Hay Quality

There are many factors, all variable, that can affect the quality of hay. Some of these are listed on page 134.

Did You Know?

An average goat will consume 490 pounds of hay during a typical 90-day winter feeding period.

HOW MUCH IS THAT BALE OF HAY?

Most hay dealers sell hay by the ton instead of the bale. Purchasing hay by the ton allows you to know precisely how much hay you're getting for your money, provided the hay is cured properly and accurately weighed.

To convert the price per ton to price per bale, determine in pounds the average weight of bales you're purchasing (most bales of alfalfa hay weigh from 65 to 75 pounds each), then divide the price per ton by 2,000 and multiply the result by the average weight of the bales. This results in the price per bale.

Stage of Maturity at Time of Harvest

Alfalfa should be harvested in the bud stage (when it has buds at the tip of its stems) or slightly later, in early bloom (when it has some purple flower petals in it and its stems are somewhat heavier than at bud stage). Alfalfa cut in full bloom or later will have fewer leaves, its stems will be coarse and woody, and it may have seedpods in it. Leafiness and soft, pliable stems are excellent indicators of quality alfalfa hay.

Clover should be harvested at the 20 percent bloom stage; it should have very few blossoms.

Orchardgrass, tall fescue, and reed canarygrass should be harvested in the boot (when grass heads are still enclosed by the sheath of the uppermost leaf and no seed heads are showing) to early-heading (when seed heads are just beginning to emerge) stages. Timothy and brome should be cut in early bloom (when the plant's tiny flowers are beginning to form in the uppermost seed head) and when fully headed. Early-cut grass hay is greener than late-cut; plump brown seeds are indicators of less nutritious, fully matured grass hay.

Grass and legume mixes should be cut when their legume component reaches the ideal stage for harvest.

Weediness

Goats appreciate most weeds, so weediness doesn't affect the quality of goat hay to the degree it does for monograstric species such as horses. Goat hay, however, should be free of poisonous plant species; invasive weeds such as thistle and jimsonweed, which can easily spread to your property via baled hay; and eye-pokers such as blackberry canes and hawthorn twigs.

Storage

A great deal of good hay is ruined through improper storage. Choose small bales stored indoors, forgoing bottom bales that harbor mold and top bales that are splotched with bird (and sometimes cat) droppings.

Large bales kept under cover are a better buy than field-stored bales, which, in areas of heavy rain- or snowfall, can represent up to 35 percent

HAY-QUALITY CHARACTERISTICS

- Excellent hay: Alfalfa cut in late bud to early bloom; clover at 20 percent bloom; grasses cut in the boot stage. Hay is bright to medium green, leafy, and free of dust, mustiness, and mold. Hay of this quality is high in protein, energy, minerals, and carotene, but relatively low in fiber. For goats, legumes of this quality should be fed very judiciously (if at all); however, high-quality grass hay is ideal for kids, lactating does, and other goats requiring palatable, high-energy forage.
- Good hay: Legumes cut in 50 percent bloom; grass cut as it begins to head. This hay is reasonably soft, leafy, green, and free of dust, mold, and mustiness.
- Fair hay: Legumes or grasses harvested at full bloom. This hay is tinged green or yellow. It's stemmy and low in protein, energy, minerals, and carotene and high in fiber. It may contain a moderate amount of dust — don't feed it inside an enclosed barn. This hay is usable but should be supplemented with grain or protein blocks.
- Poor hay: Any legume or grass hay cut after full bloom. It's stemmy, has few leaves, and is yellow or brown. It may be dusty, moldy, or musty. This type of hay is a poor buy; it's nutritionally bereft, it may contain harmful amounts of dust or airborne mold, and goats will waste plenty of it. Buy a better grade of hay; you'll be glad you did.

waste. If stored outside, bales should be placed on pallets or poles to get them up off the ground and covered with plastic tarps after they're fully cured (covering too soon or when they're damp from dew or rain results in spoiled hay). Don't store bales close together, abutting each other from the sides (instead store them in a line, end to end), and don't stack them; both practices trap moisture between the bales and ultimately ruin the material.

Mineral Supplements to Round Out the Mix

To prevent vitamin and mineral imbalances, responsible goat producers provide their animals with loose or lick-type mineral supplements appropriate to their locale, the class of goats consuming the product, and the other feed the goats are eating. As is true of every other aspect of goat nutrition, mineral supplementation is a very complex issue and a one-size-fits-all approach simply won't work. The best way to arrive at a supplement that works for your animals is to discuss your needs and your feeding program with your county Extension agent, with a goat nutritionist associated with your state university's agriculture college, or with the company that manufactures your bagged feed.

10

Keeping
Goats Healthy
(and What to Do
When They Aren't)

WHEN IT COMES TO YOUR GOATS' HEALTH, because I'm not a veterinarian, I've made no attempt to include specific treatment protocols here. Your first source of such information should be a knowledgeable veterinarian, with your experienced goat mentor(s) as sources of information and advice. If none of these individuals is available, until you can consult one, a great deal of useful material is available on the Internet (see Resources).

As we've seen in the earlier chapters of this book, the trick to having healthy goats is to start with healthy foundation stock, then to manage those goats as carefully as you can. It's fairly simple to keep your herd in the pink if you nip health problems in the bud. To do this, incorporate these strategies into your practice.

- Monitor your goats at least once a day; more often is even better. Count noses (sick goats tend to drift away from the main herd) and make certain all goats appear to be healthy (refer to the Healthy Goat, Sick Goat chart in chapter 5) and uninjured. Address sickness or injuries immediately —

don't wait to see if animals get better by themselves. If you don't know what's wrong with a goat or you're not positive you know how to treat what ails the animal, don't wing it; *get help*.

- Maintain clean, dry surroundings for your goats. Provide adequate drainage in barn lots and shelter areas; regularly shovel manure out of structures; and provide adequate, draft-free shelter from the elements.
- Don't keep too many goats for your facilities. Overcrowding, especially in housing areas, leads to stress as well as manure buildup that contributes to contamination by a number of infectious pathogens.
- Police your pastures and goat yards to remove accidents waiting to happen: protruding nails, pulled-down wire, sharp edges on metal buildings, hornets' nests, and so on. Set up a weekly or monthly schedule to check all goat areas and follow it. Don't allow goats to climb on stored hay; they can and do fall between the bales and get stuck, in the process injuring themselves or even suffocating.
- Provide adequate predator protection. Maintain truly predator-proof fencing (though this is difficult and in some locales impossible), house goats close to human habitation from twilight through mid-morning, and/or add livestock guardian animals to each herd's mix.
- Don't unduly stress your goats during handling, hauling, or weaning. Stress is a killer; prevent its occurrence whenever you can.
- Assemble a first-class first-aid kit and learn how to use it (see box on page 140). Keep the kit where you can find it in an emergency. Whenever you use an item from the kit, replace it right away. Perform an inventory of the kit every year, checking expiration dates, and discard outdated products.
- Closely monitor pregnant does; be with them at kidding time. Know how to assist if necessary and keep a well-stocked kidding kit on hand.
- Choose goat-friendly feeds; don't improvise. If you don't know how to formulate a safe, nourishing goat concentrate, hire a nutritionist to do it for you or feed commercially bagged goat concentrates. Feed quality hay, avoiding dusty and moldy bales. Make certain timid goats aren't being pushed away from feeders and hay bunks; shy goats need their share too. Make all feed changes gradually, allowing goats time to get used to new types or quantities of feed.
- Control rats and mice in feed areas. Don't allow barn cats, dogs, or poultry to defecate in hay and other feeds.
- Provide copious supplies of fresh, clean water. If you wouldn't drink it, neither will fastidious goats. Add tank and bucket heaters in the winter

and place watering facilities in the shade during the hot summer months; you want your goats (especially bucks, wethers, and lactating does) to drink as much water as they can.

■ Quarantine all incoming animals and thoroughly disinfect your quarantine pen after each occupation. Make no exceptions, even for homecoming show goats. Maintain a hospital pen for sick goats (use your quarantine pen if it isn't already occupied); don't leave them with the rest of the herd.

■ Have dead animals, especially aborted fetuses, tested to determine cause of death, and ask your vet for details after test results are in. Remove dead animals and birthing tissues immediately (don't allow livestock guardian dogs to eat them). If they aren't going to be necropsied or tested for pathogens, properly dispose of them by burning, burial, or composting.

■ Vaccinate for enterotoxemia and tetanus no matter where you live. Consider testing for caseous lymphadenitis and adding the appropriate vaccine (autogenous — that is, made from organisms from a specific disease outbreak — or over the counter) to your vaccination regimen. (See Caprine Maladies You Should Recognize, on page 144, for more information on these illnesses.) Discuss a vaccination program with your vet and add any other immunizations recommended for goats raised in your area.

■ Have your goats tested for caprine arthritis encephalitis (CAE) and Johne's disease (see Caprine Maladies You Should Recognize, on page 144) and either eliminate carriers or discuss eradication programs with your vet.

■ Trim hooves on a regular basis to maintain soundness and prevent hoof deformities.

To Be or Not to Be Your Own Vet?

Unfortunately, it's difficult (if not impossible) to find a goat-savvy vet in most parts of North America. Because of this, producers tend to be their own vets. This is not without peril, however.

Quite a few serious goat maladies closely resemble one another: for example, pregnancy toxemia and milk fever, listeriosis and goat polio, tetanus and rabies, enterotoxemia and bloat. Treating for one when your goat has the other can be catastrophic for the sick goat.

A single life-threatening symptom can have many causes. For instance, watery scours can be caused by coccidiosis, poisoning, or severe roundworm overload. You'll need to have a fecal sample tested to know for sure.

Many of the pharmaceuticals required to treat your own animals are prescription drugs, and many of these are off-label for goats (meaning that their use for treating goats is not specified on the label and is technically illegal except under the direction of a veterinarian). Most vets won't hand them out without seeing your animals. Other vets dispense prescription drugs only to established clients, but one way or another, you'll have to get them through a vet.

The most workable solution: Find a vet who's qualified to treat your goats (see Finding and Working with a Goat-Savvy Veterinarian, on page 8, for some useful tips) and establish a relationship with him or her, but also learn to address minor problems and routine veterinary procedures yourself.

Learning the Ropes

Keeping in mind that it's illegal to diagnose or prescribe without a license, find an experienced goat producer who's willing to show you how he or she handles problems in the herd. Remember that the producer needn't raise meat goats; what works for dairy and fiber goats works for meat goats too — and some dairy and fiber producers have been in the goat business for 25 years or more.

Another avenue: Join goat-oriented e-mail groups where you can read how others handle health crises and where you can post questions. The biggest, the oldest, and the best by far is The Boer Goat. You needn't own Boers to benefit from participating in this knowledgeable, friendly listserv.

And for emergencies when the vet (or your mentor) can't be reached and you don't have time to post to a listserv, phone a Goat 911 responder for free advice (see box below).

GOAT 911

Over the years, the scores of seasoned goat owners who donate their time and expertise to Goat 911 have successfully talked beginners through every conceivable caprine crisis. Print out the Goat 911 responders' list and slip it into your phone book so it's close at hand when you're rushed and need it *right now*. Find it at www.goatworld.com.

Checking Vital Signs

Whether you contact your vet, your mentor, or Goat 911 or post to a favorite e-mail listserv, be ready to provide the sick goat's vital signs (temperature, heart rate, and respiration) and to describe its symptoms in detail. These are normal values for adult goats (kids' values run slightly higher):

- Temperature: 101.5–103.5°F (38.6–39.7°C)
- Heart rate: 70–90 beats per minute
- Respiration rate: 12–20 breaths per minute

BUILD A BETTER FIRST-AID KIT

We keep our farm-based, small-ruminant first-aid kit (we use it for our sheep as well) in two five-gallon plastic food service buckets fitted with snug lids. On the top and sides we've affixed big crosses using red duct tape to indicate their contents and so the buckets are easy to spot when we need them. One bucket is further marked with SHEEP and GOATS in large letters to differentiate it from our equine first-aid kit (it's labeled HORSES). We keep the buckets in the house in a walk-in closet, and they're returned to their place immediately after each use — no exceptions.

One bucket holds emergency equipment used for the goats, sheep, and horses. It contains lead ropes and half a dozen halters ranging from our smallest small-ruminant halter (actually, an alpaca halter, but it neatly fits kids and our miniature sheep) on up to a huge one that fits our Thoroughbred mare. The horse halters are hand-tied rope versions to conserve space. When we need something in an emergency, we carry the bucket to the site and simply dump everything on the ground so we can pick out whatever we need. The bucket also contains a fencing tool and a small length of aluminum electric-fence wire for making impromptu repairs if an animal must be extracted from a fence.

The second bucket is organized with zip-top bags in several sizes. Each bag is labeled (in big black letters) according to its contents' basic uses: one bag contains wound cleanup and bandaging materials, including gauze sponges, Telfa pads, three rolls of VetWrap self-stick disposable bandage, a roll of sterile gauze bandage (2½ inches

Keep in mind that external conditions can affect your readings. A goat's temperature rises slightly as the day progresses and may be up to a full degree higher on hot, sultry days. Extreme heat and fear or anger elevate pulse and respiration. Take these caveats into consideration before you react; slightly elevated readings are sometimes the norm.

Temperature

The first thing the vet or mentor will ask is "What is the goat's temperature?" Elevated temperatures are usually associated with infections or dehydration;

wide), rolls of adhesive tape (1 inch and 2 inches wide), a partial roll of duct tape (with ¾ to 1 inch of tape left on it), two heavy-duty sanitary napkins (they can't be beat for applying pressure to stanch bleeding), a sandwich bag of flour (it works better than commercial powder to stop bleeding), a small bottle each of Betadine scrub and regular Betadine (or a similar microbicide), and a 12-ounce bottle of saline solution.

Another bag holds hardware: blunt-tipped bandage scissors, a hemostat (it can do the job more easily than tweezers), a flashlight (the flat kind you can hold between your teeth), a stethoscope, and a digital thermometer in a hard-shell case.

A third bag contains basic medicines such as our wound treatments of choice (Neosporin antiseptic ointment, emu oil, and Schreiner's Herbal Solution), topical antibiotic eye ointment, and a full tube of Probios probiotic paste.

Packed among the bags are a long package of roll cotton and two flat wooden paint stirrers from the hardware store — combined with the VetWrap, these provide everything you need to splint a broken leg.

Along with the first-aid kit, we store at the ready over-the-counter and prescription drugs we'd need in an emergency. These are located in a separate, easy-to-grab plastic basket in our pharmaceutical refrigerator (a dorm-size model) we keep in our house (though hooking it up in the barn is even better).

We keep a separate, scaled-down first-aid kit behind the seat of the truck (see Build a Traveling First-Aid Kit, page 98).

a subnormal temperature can mean hypothermia or hypocalcemia (milk fever) or indicate that a goat is dying. Both high and subnormal temperatures should be considered red flags; a goat suffering from either needs treatment right away.

To take a goat's temperature, you'll need a rectal thermometer. Veterinary models are best, but a digital rectal thermometer designed for humans works too. Traditional veterinary thermometers are made of glass and have a ring on the end to which you can attach a string. Add an alligator clamp to one end of a 14-inch length of cord and knot the other to the thermometer. This allows you to attach the clamp to the hair of the patient's tail before inserting the thermometer, which will prevent you from either losing the thermometer inside the goat (yes, it can happen) or dropping and breaking it. Glass thermometers must be shaken down after every use: hold the thermometer firmly and shake it in a slinging motion to force the mercury back down into the bulb.

Because a digital thermometer is faster, it beeps when done, and it needn't be shaken down (simply press a button and it's reset), it's the best choice for working with livestock.

Take a goat's temperature in three easy steps.

1. Restrain the goat. It's easiest to place a small kid facedown across your lap, but you'll have to halter and tie a larger animal or secure it in a grooming or milking stand. If you can recruit a helper to hold on to the goat instead of tying it up, so much the better.

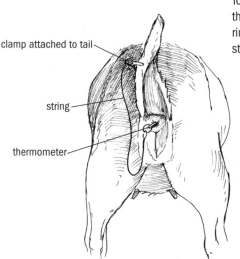

To take a goat's temperature, use a rectal thermometer with a string attached to the ring at its end. For added safety, clip the string to the goat's tail.

clamp attached to tail

string

thermometer

2. Insert the business end of the lubricated (KY Jelly, Vaseline, and mineral oil are excellent lubricants) thermometer about two inches into the goat's rectum. Don't grasp the goat's tail to facilitate insertion; goats don't like to have their tails handled and doing so is apt to trigger a rebellion.

3. Hold a glass thermometer in place for at least two minutes and a digital model until it beeps.

After recording the reading, shake down the mercury (in glass models), clean the thermometer with an alcohol wipe, and return it to its case. Always store a thermometer at room temperature.

Heart Rate

The easiest way to check a goat's heart rate is with a stethoscope: place the stethoscope on the left side of the chest wall until you hear or feel the heartbeats. You'll hear two different sounds; count the number of louder first sounds for one minute. A goat's normal heart rate is 70–80 beats per minute. If you don't have a stethoscope, take the goat's pulse by lightly pressing two fingers against the large artery on the inside of either rear leg near the groin. Count the number of pulses in 15 seconds, then multiply that number by 4.

Respiration

Simply watch the animal's rib cage and count the number of breaths it takes in one minute.

To check a goat's heart rate, place the stethoscope on the left side of the chest wall until you hear the heatbeats.

Caprine Maladies You Should Recognize

The following is a compendium of goat illnesses, complete with symptoms, treatment, prevention, and resources for each. It is meant as a general guide; as mentioned at the start of this chapter, always consult your veterinarian or goat mentors for specific diagnoses and treatment of your animals. For more information on specific conditions, see Resources.

Bloat

Bloat occurs when goats gorge on grain (perhaps through raiding an unlocked feed room), legume hay (when they aren't accustomed to eating it), or tender, high-moisture spring grass. Gas becomes trapped in the rumen and expands until it presses so hard against the animal's diaphragm that it will suffocate the animal without immediate treatment.

Symptoms: Bulging, taut sides; goat kicks at its abdomen, grunts, cries out in pain, grinds its teeth; labored breathing; collapse.
Treatment: This is an emergency situation; call your vet or mentor without delay.
Prevention: Store grain and legume hay where goats can't overindulge; feed grass hay in the morning before turning goats onto lush spring pasture.

Bottle Jaw

Bottle jaw is marked edema of the lower face and jaw. It's most noticeable in the evening, after a day of grazing, and is a last-call symptom of debilitating internal parasite infestation and severe anemia. *Note:* Bottle jaw should not be confused with milk goiter (milk neck), a harmless condition commonly seen in kids, especially Boers.

Treatment: Seek treatment without delay! Take a fecal sample to your vet for analysis and deworm according to the vet's recommendations. A good iron supplement should be added to the goat's diet (many producers swear by Geritol), as well.
Prevention: Implementation of an effective deworming program will prevent bottle jaw.

Brucellosis

Brucellosis is a serious, federally reportable disease. Called Bang's disease in cattle, *Brucella ovis (B. ovis)* in sheep, and undulant fever in humans, brucellosis is caused by bacteria from the genus *Brucella*. It triggers spontaneous abortion, retained placenta, intermittent fever, and sometimes manifests in bucks as orchitis (inflamed testicles), in does as mastitis, and as stiff, swollen joints in goats of both sexes. Brucellosis (as undulant fever) can be passed to humans who handle aborted fetuses or consume unpasteurized milk from infected livestock.

Note: While there is a blood test for brucellosis (and it's still required for interstate shipment into a few states), it sometimes gives false readings; any goat that tests positive should be retested.

Symptoms: Spontaneous late-term abortion, retained placenta, impaired fertility, intermittent fever, and joint swelling
Treatment: Except under rare circumstances, state and federal laws require the slaughter of affected animals.
Prevention: Effective brucellosis vaccine is available, but it's off-label for goats; check with your vet before using it.

Caprine Arthritis Encephalitis (CAE)

CAE is an incurable viral infection caused by a retrovirus similar to the one that causes HIV in humans. Although there is an uncommon juvenile-onset form that causes seizures and paralysis in infected kids, CAE is primarily a wasting disease of adult goats.

Note: There are several effective serology (blood) tests for CAE.

Symptoms: Swollen knees, progressive lameness; unexpected, severe weight loss; hard udder; congested lungs (CAE goats eventually succumb to chronic, progressive pneumonia).
Treatment: None.
Prevention: Because CAE is primarily passed from does to their kids via colostrum and milk, as part of a CAE-eradication program kids are removed from their dams at birth and fostered on CAE-free does or bottle-raised on milk replacer or CAE-free pasteurized milk. CAE goats and CAE-free animals should be maintained in separate herds.

CAE IN MEAT GOAT HERDS

Most dairy and packgoat breeders feel this is the most significant health problem in goats. Because market goats go to slaughter before CAE has time to manifest, most commercial producers don't share their concern. Anyone raising breeding stock, however, should consider buying from certified CAE-free herds or having their own goats tested, then implementing a CAE-eradication program.

Caseous Lymphadenitis (CL or CLA)

Caseous lymphadenitis (also sometimes called cheesy gland) is caused by the bacterium *Corynebacterium pseudotuberculosis*. It manifests as thick-walled, cool-to-the-touch abscesses containing odorless, greenish white, cheesy-textured pus. CL abscesses form on lymph nodes and lymphoid tissue, particularly on the neck, chest, and flanks, but also internally on the spinal cord and in the lungs, liver, abdominal cavity, kidneys, spleen, and brain. CL is contagious and incurable. Transmission is via pus from ruptured abscesses.

Symptoms: External abscesses; goats with internal abscesses may waste away, depending on which organs are involved.

Treatment: Any goat with a ripening abscess should be quarantined and the abscess carefully drained and treated according to a goat-savvy veterinarian's instructions. Do not allow drained pus to contaminate the goat's surroundings. Because CL is transmissible to humans, it's important to wear protective clothing. When the procedure is completed, sterilize your clothing or burn it

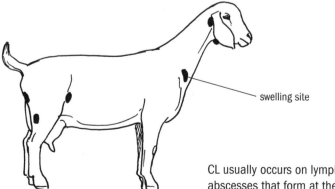

swelling site

CL usually occurs on lymph nodes. Suspect abscesses that form at these locations.

NOT EVERY ABSCESS IS CL

Don't automatically panic when a goat develops a lump; very few of them are caseous lymphadenitis. Goats are notorious for developing injection-site nodules, and these mimic developing abscesses to a T. Abscesses can also form when any of hundreds of organisms breach the skin via puncture wounds, splinters, and everyday cuts or abrasions. Goats develop salivary cysts along their jaw lines too. (The black areas in the illustration indicate abscess sites.) The only way to be certain it's CL is to have the contents of a suspicious lump cultured. Isolate goats with ripening (progressively softer-centered) lumps and proceed with treatment as indicated.

along with anything else contaminated by pus. The goat should remain quarantined until its abscess has healed.

Prevention: Buy only from CL-free herds. Several blood tests are available; have your herd tested, then vaccinate. CL-positive goats can be vaccinated with autogenous vaccine (custom made from bacteria isolated on your farm by using pus from one of your infected goats) to effectively suppress future abscesses. Some producers inoculate "clean" herds with Colorado Serum's Case-Bac vaccine for sheep.

Coccidiosis

Coccidiosis is a potentially fatal, highly contagious disease, particularly of young kids. Other livestock species are afflicted by coccidiosis but *Eimeria* protozoa are species-specific, meaning that goats can't catch chicken or dog coccidiosis (and vice versa). A single infected goat can shed thousands of microscopic coccidial oocysts in its droppings every day. If another goat accidentally ingests a sporulated (mature) oocyst, it can become ill a week or two later. As oocysts multiply and parasitize the gut, they destroy their new host's intestinal lining. Without immediate, aggressive treatment, most kids die; however, some develop chronic coccidiosis culminating in stunted growth.

Because coccidiosis is caused by minute intestinal protozoa of the genus *Eimeria*, it's actually considered a parasitic infection (which, for this book, means we'd discuss it just in chapter 11) rather than a disease. Yet given that coccidiosis is a major medical problem for goat ranchers everywhere, it deserves mention in this chapter as well.

Note: Goats develop a certain amount of immunity as they age. Nevertheless, many adult goats are carriers and continually shed coccidial oocysts. Neonatal scours can also be caused by maladies as diverse as *E. coli*, enterotoxemia, and heavy worm infestation. If you aren't sure what you're dealing with, see a vet.

Symptoms: Watery diarrhea (scouring), sometimes containing blood or mucus; listlessness; poor appetite; abdominal pain.

Treatment: Dewormers don't kill coccidia; sulfa drugs are usually the treatment of choice. Banamine can be used to calm the gut and reduce fever, while electrolytes such as Re-Sorb, Pedialyte, and Gatorade should be given orally to rehydrate stricken kids. Consult your vet or mentor for particulars; this is a life-threatening illness, so don't delay.

Prevention: Concentrates and vitamin-mineral supplements containing monensin (Rumensin) or decoquinate (Deccox) effectively quash coccidiosis before it begins; however, ingestion of Rumensin has been implicated in numerous equine fatalities, so never expose horses, ponies, donkeys, or mules to products containing it. Limit goats' exposure to oocysts by eliminating crowded conditions, cleaning pens daily, and preventing unnecessary fecal contamination by using keyhole-accessed feeders and watering stations.

Enterotoxemia

Also known as entero, overeating disease, and pulpy kidney, enterotoxemia is caused by common bacteria found in relatively small numbers in manure, soil, and even the rumens of perfectly healthy goats. When these bacteria quickly proliferate for one reason or another (overeating on grain or milk, abrupt change in quantity or type of feed, drastic weather change), they produce toxins that can kill a goat in hours. Enterotoxemia is a major killer of unvaccinated goats, particularly young kids.

Symptoms: Bloating; rocking-horse stance; goat grinds its teeth and cries out in extreme pain; seizures, foaming at the mouth, coma, death. Note that because death comes so rapidly, many times you don't have an opportunity to observe symptoms but instead may find that a kid that was the largest and seemingly strongest is dead.

Treatment: Treatment is usually ineffective because death occurs so quickly.

Prevention: Goats are prone to two types of enterotoxemia, *Clostridium perfringens* type C (which quickly kills young kids) and type D (which affects slightly older, fast-growing kids and adult goats). A single vaccine is used to treat both. The most common is CD/T toxoid (*T* is for tetanus; this three-way shot also prevents that potentially fatal disease), which is also available in an eight-way vaccine sold as Covexin 8. CD antitoxin provides short-term immunity to previously unvaccinated goats.

Floppy Kid Syndrome

The jury is still out regarding the precise cause or causes of floppy kid syndrome. What is known is that since 1987, this baffling syndrome has occurred with frightening regularity in herds of all types and breeds, but especially in meat goat kids.

Symptoms: Afflicted kids appear normal at birth but develop profound muscular weakness (flaccid paralysis) at three to 21 days of age. Kids are initially lethargic and quickly develop a wobbly gait until they are soon unable to stand. They lose all muscular control and become totally flaccid as the syndrome progresses. Because they cannot suckle, they quickly become dehydrated and must be rehydrated via a stomach tube or other means.

Treatment: This varies widely. Some kids recover with or without additional treatment but the overall mortality is high. Relapses are common. Consult your vet or county Extension agent for up-to-date information about this devastating syndrome.

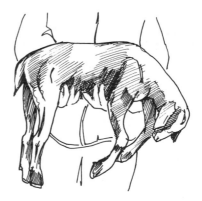

At the height of floppy kid syndrome, afflicted kids are totally limp; none of their muscles seems to work.

Goat Polio

Goat polio (also called polioen-
cephalomalacia or cerebrocorti-
cal necrosis) is in no way related
to the viral disease called polio
(poliomyelitis) in humans. Polio-
encephalomalacia is a neurological disease caused by a thiamine
(B_1) deficiency that culminates
in brain swelling and the death
of brain tissue.

Stargazing is a symptom of goat polio.

Symptoms: Disorientation, depression; stargazing, staggering, weaving, circling, tremors; diarrhea; apparent blindness; convulsions; death in one to
three days.

Treatment: If treatment (thiamine injections) begins early enough, affected
goats begin improving as soon as in two hours. Thiamine is a prescription
medicine; you must obtain it from your vet.

Prevention: Thiamine deficiencies can be triggered by eating moldy hay or
grain; overdosing with amprollium (CoRid) when treating for coccidiosis;
ingestion of certain toxic plants; reactions to dewormers; and sudden changes
in diet, including weaning. Overuse of antibiotics can contribute to thiamine
deficiencies too.

Hoof Rot

Hoof rot (sometimes called foot rot) is caused by an interaction of two bacteria, *Bacteroides nodosus* and *Fusobacterium necophorum*. *F. necophorum* is commonly present in manure and soil wherever goats or sheep are kept; it's when
this bacterium forms a synergistic partnership with *B. nodosus* that hoof rot
occurs. *F. necophorum* can live in soil for only two to three weeks, but it can
live in an infected hoof for many months. If it's not introduced to your herd
via infected goats or sheep, you'll never have foot rot, so it's vitally important
you not buy infected animals.

Symptoms: Goats with hoof rot are excruciatingly lame; they may hold up an
infected foot and hop on three legs; if one or both forefeet are infected, they
consistently kneel to feed. Trimming infected hooves exposes a putrid-smell-

ing pasty substance lodged between the horny outer surface of the hoof and its softer inner tissues. You'll definitely recognize hoof rot by its stench.

Treatment: The infection is spread from infected hooves to soil to healthy goats and sheep, so isolate all infected animals (make no exceptions). To expose disease-causing bacteria to oxygen, trim back the infected hooves to the affected areas and remove as much "rot" as you possibly can. Treat according to your veterinarian's directions; treatments include the use of topicals (Coppertox, merthiolate, mercurochrome, bleach solution), foot baths and soaks (zinc sulfate, copper sulfate, formaldehyde), and antibiotics.

Prevention: Don't buy goats or sheep from sales barns or from infected herds. Make certain commercial transporters who carry your animals thoroughly disinfect their trailers after every trip. Disinfect your shoes after visiting infected facilities. Because hoof rot flourishes in manure and muck, keep barnyards and holding areas as dry as you can. Quarantine infected animals and set up a footbath (ask your vet which chemicals to use and at what strength they should be mixed) for noninfected stock where they must walk through it at least twice a day.

Hypocalcemia (Milk Fever)

Hypocalcemia is caused by a drop in blood calcium a few weeks prior to and immediately after kidding and is easily confused with pregnancy toxemia. Both are life-threatening conditions; call your vet or goat mentor without delay.

Symptoms: Does seem depressed, losing interest in eating and becoming progressively weaker until they lie down and won't get up again; mild bloat; subnormal temperature as the condition progresses.

Treatment: Proceed according to your vet's recommendations. The usual treatments include oral dosing with energy boosters such as glucose, NutriDrench, and Magic and the use of oral or injectable calcium substances such as calcium gluconate and CMPK (a fluid calcium, magnesium, phosphorus, and potassium product).

Prevention: Provide pregnant does with a 2:1 calcium-to-phosphorous mineral mix at all times; feed alfalfa hay or pellets to late-gestation and lactating does; feed five or six Tums to late-gestation and lactating does every day.

Johne's Disease

Johne's (pronounced YO-neez) disease, also called paratuberculosis, is a contagious, slow-developing, progressively fatal disease of cattle, sheep, goats, deer, antelope, and bison. It's most commonly seen among dairy cattle. Johne's disease is caused by *Mycobacterium paratuberculosis*, a close relative of the bacterium that causes tuberculosis in humans, cattle, and birds. The disease is rampant worldwide; according to Johne's Information Center statistics, 7.8 percent of America's beef herds and 22 percent of our dairy herds are infected with *M. paratuberculosis*. Johne's disease typically enters goat herds when an infected but healthy-looking animal is added to the mix. The infection then spreads to other animals (goats — generally young kids — become infected through oral contact with contaminated manure from an infected goat). Kids are also infected by nursing from infected dams.

Symptoms: Progressive loss of condition; weakness.
Treatment: None.
Prevention: Adults should be tested and the herd divided into infected and disease-free groups, then maintained separately from one another. Kids from infected does must be removed from their dams before they nurse and must be grafted onto disease-free foster mothers or bottle-raised using milk replacer or milk known to be free of *M. paratuberculosis*.

Listeriosis

Listeriosis is caused by a common bacterium called *Listeria monocytogenes*, found in soil, plant litter, water, and even healthy goats' guts. There are two types of listeriosis in goats: one causing paralysis and the other (and more common) type leading to encephalitis (inflammation of the brain). Problems begin when dramatic changes in feed or weather conditions occur, causing bacteria in the gut to multiply. Parasitism and advanced pregnancy can trigger bacteria proliferation too.

Note: Humans can contract listeriosis; take due precautions when handling afflicted goats.

Symptoms: Disorientation, depression; stargazing, staggering, weaving, circling; one-sided facial paralysis, drooling; rigid neck with head pulled back toward flank. Symptoms resemble goat polio, rabies, and tetanus, so call in

A goat with a rigid neck and its head pulled back tight toward its flank is likely suffering from listeriosis or tetanus.

your vet to make certain you know what you're dealing with, especially if you suspect rabies.

Treatment: Treat according to vet's recommendations.

Prevention: Avoid drastic changes in type and amount of feed; never feed moldy hay, silage, or grain.

Mastitis

Mastitis is inflammation of the udder and can be caused by a number of bacterial and staph agents. It can be triggered by substandard milking hygiene and delayed milking in dairy goats; udder injuries; stress; and milk buildup after meat does wean their kids. Untreated, it can progress to gangrene mastitis and death.

Symptoms: Swollen, hot, hard udder; extreme pain; lameness; loss of appetite; fever; decreased milk production; clumps, strings, or blood in the milk; watery-looking milk. Gangrene mastitis: bruised-looking, extremely swollen, and painful udder that turns blue as infection takes hold.

Treatment: Both intramammary infusions of antibiotics and systemic antibiotics are generally used. It's best to have milk samples cultured to determine which bacteria are involved and thus which medications to use.

Prevention: When milking does, practice good milking sanitation. Reduce lactating does' grain rations for several successive days before weaning and switch them from legume to grass hay, then eliminate all grain post-weaning until does' udders have dried up.

Pinkeye

Pinkeye, technically known as infectious keratoconjunctivitis, is a bacterial eye infection commonly spread by flies and dust. Because it is contagious and can cause ulceration of the cornea and permanent blindness, it should be treated aggressively.

Symptoms: Watery eye, cloudy cornea, light sensitivity.
Treatment: Isolate infected goats in shade or a darkened building. Many over-the-counter products effectively eliminate pinkeye; ask your vet for recommendations. Antibiotics are sometimes used to treat the infection.
Prevention: The only prevention for animals who have not yet contracted pinkeye is to keep them away from infected animals.

Pneumonia

Pneumonia, a serious and potentially fatal inflammation or infection of the lungs, is caused by any of a host of bacteria, fungi, and viruses that often gain a toehold due to environmental factors such as stress; aspiration of milk, vomit, or drench material; damage caused by lungworm infestation; exposure to drafts or being hauled in open goat cages; and dusty feed, bedding, or surroundings. Some forms are contagious.

Symptoms: Loss of appetite, depression; rapid respiration and labored breathing; goat stands with forelegs braced wide and neck extended; thick, yellow nasal discharge; congestion, coughing, an audible "rattle" in the chest.
Treatment: Pneumonia is a red-alert emergency; call your vet!
Prevention: Remove environmental factors. Many goat producers vaccinate for pneumonia using off-label cattle vaccines; discuss this with your vet.

Pregnancy Toxemia and Ketosis

Pregnancy toxemia and ketosis are potentially life-threatening metabolic disorders that afflict does during their final weeks of pregnancy (pregnancy toxemia) and less commonly, the first week or so after kidding (ketosis). They mainly occur in obese or thin does and does carrying multiple kids, but can strike any doe that isn't fed properly during the last few weeks of gestation and the period immediately after kidding. Genetics appear to be a factor and Boer does seem especially prone to pregnancy toxemia, so Boer owners should know how to recognize these disorders and keep on hand the materials needed to treat them.

Pregnancy toxemia/ketosis is caused by excess ketones (chemical substances that the body makes when it does not have enough insulin in the blood). These accumulate when a late-gestation doe ingests fewer calories than she needs to support herself and her fetuses or when a milking doe produces more milk than her energy intake can provide for. In both cases the doe's body mobilizes body fat reserves to fulfill her need for carbohydrates.

A good way to check a suspect doe is to catch her urine in a container (fasten a clean tin can to a broom handle and wait for her to get up; does usually relieve themselves on arising) and test it using ketone strips available at any drug store.

Symptoms: Lack of appetite, depression, unwillingness to stand, muscle tremors, staggering, teeth grinding and other indications of pain; eventually coma and death.

Treatment: If you've never dealt with this before and one of your does shows early symptoms, call your mentor or vet without delay! Treatment must be swift and aggressive. She must be treated around the clock with glucose/sugar supplements in the form of a commercial NutriDrench-type product, Magic (two parts light corn syrup to one part molasses and one part corn oil; see page 167), propylene glycol, oral dextrose, or intravenous glucose. She will also require exercise (get her up and moving), B-complex injections to stimulate her appetite, and probiotics to further stimulate her appetite and keep her rumen functioning. If she survives to kidding, expect to help deliver her kids as she will probably be too weak to do it by herself.

Prevention: Does should be neither thin nor obese going into pregnancy. Don't overfeed in early pregnancy but do increase carbohydrate (energy) intake during a doe's last trimester by adding judicious amounts of grain and/or alfalfa to her diet. Make sure does get enough exercise (feeding and watering at opposite ends of their pasture or enclosure encourages movement). After kidding, heavily-milking does require additional high-energy feed as well.

Scrapie

You'll probably never see a case of scrapie (pronounced SCRAPE-ee) but you may be required to tag your goats' ears with scrapie tags, so it's best to understand what the disease involves. Scrapie is a transmissible spongiform encephalopathy (TSE) of sheep and goats similar to bovine spongiform encephalopathy (better known as BSE or mad cow disease). It's a progressive, fatal disease that systematically destroys the central nervous system. Symptoms appear when infected goats are two to five years old; they die one to six months later. What causes scrapie is not fully understood: genetics are part of the picture, but there's also an as-yet-unidentified infective agent involved.

Note: Many classes of goats are required to be ear-tagged and identified as part of the USDA's mandatory scrapie-eradication program. Because program requirements are constantly changing, contact your state Animal and Plant Health Inspection Service (APHIS) representative for current information. You can locate this office via the APHIS Veterinary Services Area Offices web page at www.aphis.usda.gov.

Scrapie tags must be purchased from vendors licensed by the USDA. Ask your APHIS representative for details.

Symptoms: Progressive wasting; tremors, stumbling; excessive salivation; itchiness; blindness; coma and death.
Treatment: None.
Prevention: None.

Soremouth

Soremouth goes by many names, among them contagious ecthyma (CE), contagious pustular dermatitis, scabby mouth, and orf. It's caused by a parapox virus and is of special concern because humans who handle infected goats can catch it. As its name implies, soremouth raises blisters and pustules on the lips and inside the mouths of infected goats, making their mouths quite sore indeed. Infected kids go off feed and lose condition, but most recover spontaneously in one to four weeks. The virus that causes soremouth can survive up to 30 years in soil, so once it's on your farm, soremouth will be an ongoing problem.

Symptoms: Blisters and pustules on kids' lips can spread soremouth to their dams' teats; infected does may not allow their kids to nurse. As the disease progresses, pustules burst and crusty scabs form.
Treatment: Softening ointments can be applied to scabs. Soft, palatable feeds should be given to infected goats.
Prevention: A live vaccine for soremouth is available. It should not be used to vaccinate clean herds, however, because as they're shed, live vaccination scabs contaminate the herd's immediate environment so that when the next generation of kids arrives, they'll contract full-blown soremouth.

Tetanus

Tetanus occurs when wounds are infected by the bacterium *Colostridium tetani*, which thrive in anaerobic (airless) conditions such as those found in deep puncture wounds, fresh umbilical cords, and wounds caused by recent castration. Unless treated very early and aggressively, tetanus is nearly always fatal.

Symptoms: Early on: stiff gait, mild bloat, anxiety. Later: a rigid rocking-horse stance; drooling, inability to open the mouth (hence the common name for tetanus, lockjaw); head drawn hard to one side; tail and ear rigidity; seizures, then death.
Treatment: Tetanus is a red-alert emergency; call your vet without delay.
Prevention: All goats, without exception, should be vaccinated against tetanus. It's a terrible way to die.

Urinary Calculi (Urolithiasis, UC)

Urinary calculi are mineral salt crystals ("stones") that form in the urinary tract and block the urethras of male goats (though both sexes are afflicted, stones pass easily through does' relatively larger, straighter, and shorter urethras). The condition is a red-alert emergency that requires immediate medical attention. It won't correct itself, and if left untreated, the afflicted goat's bladder will burst and the goat will die.

Symptoms: Anxiety, restlessness, pawing the ground; goat grinds its teeth and cries out in pain; straining to urinate; rocking-horse or hunched-over stance; impaired flow of urine (dribbling).

Treatment: Call your vet; do not delay!

Prevention: Feed male goats a balanced 2:1 calcium–phosphorus ration. To accomplish this, feed high-quality grass hay instead of alfalfa and offer very little grain. Adding minute quantities of ammonium chloride to the diet may prevent some (but not all) types of calculi from occurring. Provide plenty of clean, palatable drinking water (add a tank heater in the winter and keep water in the shade during the hot summer months).

White Muscle Disease

White muscle disease, also called nutritional muscular dystrophy, is caused by a serious deficiency of the trace mineral selenium. Most of the land east of the Mississippi and much of the Pacific Northwest are selenium-deficient; these are the areas where white muscle disease is most likely to occur.

Note: The best source of information regarding selenium conditions in your immediate locale (especially in areas where selenium levels may vary widely from farm to farm) is your county Extension agent.

Symptoms: Kids: may be stillborn; weakness, inability to stand or suckle, tremors; stiff joints; neurological problems. Adults: infertility, abortion, difficult births, retained placentas; stiffness; weakness, lethargy.

Treatment: Determine whether you're located in a selenium-deficient area before treating goats for white muscle disease. Injections of Bo-Se (a prescription supplement you can get from your vet) often dramatically reverse symptoms, especially in neonatal kids.

Prevention: All goats raised in selenium-deficient areas should be fed selenium-fortified feeds, have free access to selenium-added minerals, or be given Bo-Se shots under a vet's direction. To prevent kidding problems and protect unborn kids, does should be injected with Bo-Se three or four weeks prior to kidding.

Stop Disease in Its Tracks — Vaccinate!

Not everyone routinely vaccinates his or her goats, but everyone should. At minimum, wise producers vaccinate for enterotoxemia and tetanus because both are common killers. While enterotoxemia vaccine (both toxoid and antitoxin) can be purchased by itself, a better buy is a combination toxoid that protects against tetanus too (it's marketed as CD/T toxoid). Enterotoxemia and tetanus protection is also available in an eight-way combination vaccine called Covexin 8, and it's a favorite of many goat producers.

Depending on where you live and what sort of "bugs" haunt your locale, your vet or goat mentor may recommend other vaccinations as well.

TOXOID OR ANTITOXIN?

- Toxoid is the vaccine to choose when you're seeking long-term immunity for your goats. It's given in stages: an initial injection followed three weeks later by a second shot, and that followed by boosters once or twice a year. Immunity, however, is not immediate, so if an unvaccinated goat becomes sick (in the case of enterotoxemia) or injured (thus in danger of contracting tetanus), *antitoxin* is injected to impart immediate immunity.
- Toxoids are available either as stand-alone tetanus or as combination CD/T vaccines. Antitoxin is one or the other — tetanus antitoxin or CD antitoxin — but the two are never combined.
- Unfortunately, tetanus and CD/T vaccine failures are not uncommon. Don't assume your animals can't get sick because they're vaccinated. While most sources recommend annual tetanus and CD/T boosters, many producers boost at six-month intervals instead.
- Kids acquire immunity from disease via their vaccinated dams' colostrum (it protects them until their own immune systems kick in, at 8 to 10 weeks of age, the time when their toxoid immunizations should begin), so producers who vaccinate usually boost pregnant does' immunity with toxoid vaccines four or five weeks prior to kidding.
- To be absolutely certain an animal is protected, tetanus antitoxin is routinely administered when a kid is castrated or when any goat suffers a deep puncture wound.
- The protection imparted by antitoxin lasts only a short time, a week or two at most. If a goat is still ill or its wound hasn't healed at the end of seven days, it will require revaccination with antitoxin.

Antibiotics: Good or Bad?

Nowadays, most people know that antibiotic overuse is a real and rapidly growing problem. Many of those entering the goat business, having bought in to the hardy, tin can–eating goat mystique, may think, "I'm not going to use antibiotics on my goats." Yet unless a producer is raising organic goats and is willing to go off the organic program with seriously ill goats or to lose some goats to disease, avoiding antibiotics simply isn't feasible.

Goats are prone to a plethora of life-threatening maladies that respond quite well to antibiotics. Unfortunately, not all pathogens respond to the same antibiotics. Those listed in the following chart are some you'll likely use if prescribed for your goat's condition by a veterinarian who has firsthand knowledge of you and your animal or as over-the-counter treatments approved for goats. There are other over-the-counter antibiotics available, but they are not approved for use in goats.

ANTIBIOTICS COMMONLY USED TO TREAT GOATS

Trade Name	Antibiotic	Over the Counter (OTC) or Prescription (RX)
Liquamysin, LA-200, Biomycin, Oxybiotic	oxytetracycline	OTC
Naxcel Sterile Powder	ceftiofur sodium	RX (off-label)
Nuflor Injectable Solution	florfenicol	RX (off-label)
Penicillin (long-acting)	penicillin G benzathine	OTC
Penicillin (regular)	penicillin G procaine	OTC

Sources: *Antibiotics Commonly Used in Livestock Production* (www.sheepandgoat.com/articles/antibiotictable.html); *Goat Drugs and Dosages* (www.goatworld.com/health/meds/dosages.shtml); *List of Diseases and Conditions in Meat Goats* (www. jackmauldin.com/diseases.htm)

Antibiotic Dos and Don'ts

While antibiotics are enormously helpful and we have come to rely on them for the treatment of many illnesses in all our animals, there are a few important guidelines to follow when using them (or deciding not to use them):

- Don't use antibiotics indiscriminately. Know what troubles each sick goat, which antibiotics are likely to be effective, and in what dosage and duration each antibiotic should be given.
- Strictly enforce suggested withdrawal times prior to slaughter so your customers don't ingest antibiotic-laced meat. Withdrawal times are also important if you consume dairy products made of milk produced by one of your does.
- Follow dosage directions to the letter. Use precisely the recommended dosage and complete the series as directed (otherwise, weak pathogens die but "super germs" survive to proliferate).
- Because antibiotics destroy good bacteria as well as bad, *always* follow antibiotic treatment with oral probiotics such as Calf Pac, Probios, and Fast-Track to restore the patient's rumen to good health.

Standard Use	Withdrawal Time prior to Slaughter	Notes
Broad-spectrum antibiotic use; pneumonia, shipping fever, foot rot, bacterial enteritis, wound infection, acute metritis pinkeye	28 days	Interferes with teeth and bone formation — not to be used for pregnant does or kids under 12 months of age.
Broad-spectrum antibiotic for respiratory ailments; especially used for pneumonia, hoof rot, *E. coli*	None	Comes in two components that must be rehydrated.
Broad-spectrum antibiotic for respiratory ailments; especially used for pneumonia and hoof rot, and orally for *E. coli* scours	28 days (intramuscular injection); 38 days (subcutaneous injection)	Because Nuflor doesn't have to be rehydrated, it keeps much longer than rehydrated Naxcel under regular refrigeration.
Shipping fever, upper respiratory infections	30 days	
Bacterial pneumonia, enterotoxemia, mastitis	8 days	Take care administering penicillin; injected into a vein, it can kill.

HELPFUL CONVERSIONS

1 milliliter (1 ml) = 15 drops = 1 cubic centimeter (1 cc)

1 teaspoon (1 tsp) = 1 gram (1 gm) = 5 cubic centimeters (5 cc)

1 tablespoon (1 tbsp) = ½ ounce (½ oz.) = 15 cubic centimeters (15 cc)

2 tablespoons (2 tbsp) = 1 ounce (1 oz) = 30 cubic centimeters (30 cc)

1 pint (1 pt) = 16 ounces (16 oz) = 480 cubic centimeters (480 cc)

Calling the Shots

Every goat producer quickly becomes adept at giving shots. Here are the steps:

1. Read the label before selecting the correct vaccine or other injectable, then read the label again after you've chosen the drug. Don't omit this very important step!

2. Use the smallest disposable syringe that will do the job; small is easier to handle than a big, bulky reusable syringe, especially for women with small hands. Disposable syringes are inexpensive and readily available and will help you avoid the work of sterilizing reusable syringes between uses. Never try to sterilize disposable syringes — boiling compromises their integrity; it simply isn't wise to reuse them.

3. Choose the correct needle for the job. Nearly all goat injections are given subcutaneously (SQ; under the skin) and should be given using 18- or 20-gauge needles, ½ or ¾ inch long. Some vets prefer that certain antibiotics be given intramuscularly (IM; into a major muscle mass). In these cases, use 18- or 20- gauge needles, 1 to 1½ inches long. Large-gauge, long transfer needles (for pulling vaccine or medicine out of the bottle) are easiest to use. Some antibiotics are very thick, and their carriers make these injections really sting. For these, some producers choose 16-gauge needles in order to inject the fluid quickly before the goat objects.

4. Select enough needles to do the job. You'll need a new needle for each goat, plus a transfer needle to stick through the rubber cap on each drug. For example, if you're vaccinating ten goats using CD/T toxoid and Super Poly Bac (off-label under a vet's direction to vaccinate against pneumonia), you'll need a dozen needles. Using a new needle each time is less painful for the goat and virtually eliminates the possibility of transmitting disease (such as CAE) via contaminated needles.

5. Secure the goat. Recruit a helper, tie the goat by its collar, use a grooming or milking stand — somehow make sure the goat is held reasonably still while you do the deed.

6. Insert a new, sterile transfer needle through the cap of each pharmaceutical bottle. As you draw each shot, attach the syringe to the transfer needle and draw the vaccine or drug into it, then detach the syringe and attach the needle you'll use to inject the pharmaceutical into the goat. *Never poke a used needle into the cap to draw vaccine or drugs!*

7. If you're drawing 3 cc of fluid from the bottle, inject 4 cc of air (to avoid the considerable hassle of drawing fluid from a vacuum), then pull a tiny bit more than 3 cc of fluid into the syringe. Attach the needle you'll use to inject the goat and then press out the excess fluid to remove any bubbles created as you drew out the vaccine or drug.

8. Select the best injection site. You'll want to inject the vaccine or drug where it will work best but without injuring the goat or damaging expensive cuts of meat. Because injection-site nodules can be mistaken for caseous lymphadenitis, many producers now inject all pharmaceuticals between the front legs, where they're less likely to be noticed and where caseous lymphadenitis abscesses don't normally occur. Other preferred sites for subcutaneous injections are the neck, over the ribs, and into the "armpit." On the rare occasions when intramuscular injections are called for, they're

PREPARING THE SYRINGE

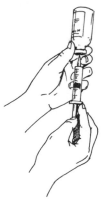

1. Use a transfer needle to withdraw shots from the bottle. Disconnect the syringe, leaving the transfer needle in place, and attach a single-use needle for the shot.

2. Inject air into the bottle to facilitate withdrawing each shot.

3. Eject a tiny bit of fluid from the needle to remove any air bubbles that may have formed.

generally given into the thick muscles of the animal's neck. *Note:* When injecting a relatively large volume of fluid, break the dose into smaller increments and inject it into more than one site (e.g., in adult goats, no more than 10 cc of penicillin should be injected into a single site).

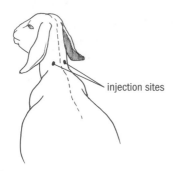

injection sites

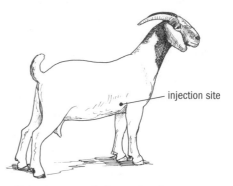

injection site

Intramuscular shots can be given into the heavy muscles of the neck to avoid damaging prime cuts of meat in slaughter goats.

Subcutaneous shots can be given in the armpit.

9. If it's remotely dirty, swab the injection site with alcohol before administering the injection (prepackaged alcohol swabs are easy to use). Never inject anything into damp, mud- or manure-encrusted skin.

10. To give a subcutaneous injection, pinch a tent of skin and slide the needle into it parallel to the goat's body. Take care not to push the needle through the tented skin and out the opposite side or to prick the muscle mass below it. Slowly depress the plunger, withdraw the needle, and rub the injection site to help distribute the drug or vaccine.

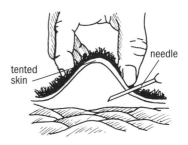

needle

tented skin

11. If you absolutely must give an intramuscular injection, here's how it's done: Quickly but smoothly insert the needle deep into muscle mass, then aspirate (pull back on) the plunger ¼-inch to see if you hit a vein. If blood rushes into the syringe, pull out the needle, taking great care not to inject any drug or vaccine as you do so, and try another injection site.

12. After injecting one product, you can use the same needle and syringe to inject the next vaccine or drug into the same goat — but only if you use a new transfer needle.

THE USE OF HEALTH CARE PRODUCTS (INCLUDING DEWORMERS)

You'll want to get the most out of the vaccines, medicines, and dewormers you buy, so choose and use them to your best advantage.

When you buy a new product, first read the label. Check and follow the storage instructions. Some drugs must be refrigerated, some must be kept out of direct light, and some should be shaken before use and aren't effective if this step isn't taken. Note any warnings or contraindications: Certain pharmaceuticals cause abortion if used on pregnant does.

Reread the label directly before using any product; it's easy to forget particulars between uses. If a product has been inadvertently stored incorrectly, discard it. Also check its expiration date before using a drug or vaccine, and pitch it if it's outdated.

Don't buy the large, economy size of any product if you can't use it up before it reaches its expiration date. Because many products are packaged for cattle, it can take a long time for your 20 goats to use it up. One alternative is to try to find another small-scale goat or sheep owner who'll split the cost and the product with you.

Epinephrine — Don't Leave Home without It

Epinephrine, also called adrenaline or epi, is a naturally occurring hormone and neurotransmitter manufactured by the adrenal glands. It was first isolated and identified in 1895 by Napoleon Cybulski, a Polish physiologist, and was artificially synthesized in 1904 by Friedrich Stolz. It's widely used to counteract the effects of anaphylactic shock, a serious and rapid allergic reaction that can kill.

Any time you give an injection to a goat, no matter the product or amount injected, you must be prepared to immediately administer epinephrine to counteract an unexpected anaphylactic reaction. If a goat goes into anaphylaxis (indicated by glassy eyes, increased salivation, sudden-onset labored breathing, disorientation, trembling, staggering, or collapse), you won't have time to race to the house to grab the epinephrine — you might not even have time to fill a syringe — you have to be ready to inject the epi *immediately*.

Many goat producers keep a dose of epinephrine in a syringe in the refrigerator. Kept in an airtight container (such as a clean glass jar with a tight-fitting lid), it will keep as long as the expiration date on the epinephrine bottle. Take it with you every time you inoculate a goat. It may save the life of a valuable or much loved animal. Standard dosage is 1 cc per 100 pounds. Be very careful not to overdose, which causes the heart to race.

Previously available over-the-counter epinephrine is now a prescription drug and is available only through a vet. An alternative some goat owners favor is over-the-counter Primatene Mist sprayed under a shocky goat's tongue. Every 14 or 15 squirts of Primatene Mist contains the same amount of epinephrine as a 1 cc dose.

Drenching a Goat

Liquid medicines, dewormers, and energy boosters are given orally as drenches that can be administered using catheter-tip syringes (not the kind you use to give shots) and even turkey basters — but the most efficient way is to use a dose syringe.

To drench a goat, restrain it (straddle its back, facing forward, if you're tall enough not to be taken for a ride) and, with one hand under its chin, elevate its head — not too high, but let gravity help you a bit. Insert the nozzle of the syringe between the goat's back teeth and its cheek (this way the goat is less likely to aspirate part of the drench) and *slowly* depress the plunger, giving the goat ample time to swallow.

A drenching syringe makes drenching goats an easy chore.

A disposable catheter-tip syringe also works.

To fill a syringe, insert the syringe tip into the bottle, as shown.

HOMEMADE MAGIC

While there are many rapidly absorbable liquid vitamin and energy products on the market (Goat Nutri-Drench is a popular one), when you need to use one on an ongoing basis, perhaps for a chronically ailing goat or a doe with pregnancy toxemia, you can whip up an inexpensive yet effective liquid energy drench right at home. It's magic! Here's how to mix and use it.

Combine:
1 part corn oil (don't substitute)
1 part molasses (unrefined or blackstrap molasses is best)
1 part Karo syrup (the clear, light kind)

Add enough dextrose (a liquid form of sugar sold at feed stores and through vet suppliers) or plain water to make the mixture easy to swallow. Store it in a cool, dark place (such as the refrigerator). It will separate but can be remixed by simply shaking it vigorously before each use.

Use Magic for does with ketosis, anemic goats (molasses is a stellar source of iron as well as a host of other vitamins and minerals), or any sick goat that seems to need a boost. Some goats drink Magic out of a bowl, but it's usually administered as a drench. Doses are generally 4 to 6 ounces administered every 8 hours but depending on its reason for use, specific dosages may vary.

To give a goat a liquid drench, slightly elevate his chin and slowly squirt the fluid between his back teeth and cheek, giving the goat ample time to swallow. Deposit pastes and semisolid medicines as far back on the tongue as the syringe comfortably reaches.

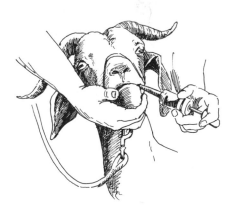

When giving a goat a semisolid drench or a medication in gel form, deposit the substance as far back on the goat's tongue as you possibly can.

In either case, keep the goat's nose slightly elevated until it visibly swallows. And be careful — if you stick your fingers between a goat's back teeth, you're likely to be bitten, and those teeth are razor sharp!

Medical Alternatives for Goats

The best way to approach alternative medicine for goats is with the help of a holistic veterinarian. To find one, contact the American Holistic Veterinary Medical Association (AHVMA) or peruse the list of member veterinarians at its website.

Holistic veterinary medicine as defined by the AHVMA is "the examination and diagnosis of an animal, considering all aspects of the animal's life and employing all of the practitioner's senses, as well as the combination of conventional and alternative (or complementary) modalities of treatment." Holistic modalities include acupuncture; Traditional Chinese Medicine; behavior modification; herbal medicine; chiropractic; meganutrient therapy; flower essences; homeopathy; and energy therapies such as Therapeutic Touch, TTouch, and Reiki. Some goats respond beautifully to alternative therapies — but one caveat should be noted: when a major malady such as pneumonia, enterotoxemia, or goat polio strikes, time is of the essence and alternative modalities invariably require time to work. It might be prudent to haul out the big guns (antibiotics) rather than lose a goat.

11

Parasites 101

INCREASING RESISTANCE TO goat anthelmintics (deworming drugs) is the most serious problem facing today's goat producers, not only here in North America but around the world as well. Goats are extremely prone to parasitism; keeping your goats from developing harmful worm loads will be one of the most difficult tasks you'll face as a goat rancher.

Goats harbor an array of internal parasites, but only a few of these can cause infestations serious enough to kill their hosts. Even so, any type of heavy worm infestation increases feed costs, decreases production and normal weight gain, stresses goats, and can contribute to life-threatening illnesses such as bottle jaw and coccidiosis (see chapter 10).

Did You Know?

Sheep and goats have the same internal and external parasites (with the exception of coccidiosis, which is species-specific). Yet the amount of dewormer needed to effectively dose each species isn't the same — in most cases, goats require considerably more product than is used to deworm sheep.

To more fully appreciate the life cycles of internal parasites, view Karin Christensen's outstanding animations at her website, The Biology of the Goat (see Resources).

You'll never vanquish worms once and for all, but you can and must control them. The best way to do this is to work with your vet to find out which parasites your goats are carrying and which dewormers are still effective on your farm.

Dewormer-Resistant Worms

In the past, experts advised sheep and goat producers to reduce worm reinfestation by deworming all of their sheep and goats at the same time, and to rotate dewormers to reduce drug resistance. Because few producers weighed their stock before deworming, many also underdosed their animals. In this manner, sheep and goats were consistently exposed to all of the available anthelmintics, often in doses too meager to effectively kill worms. Because of this, weak worms died but the strong survived. Now dewormer-resistant "super worms" have evolved to the point that most anthelmintics have lost their punch — and the drug companies aren't developing new products to replace them.

Note that while a number of dewormers can be purchased over the counter for use in cattle and sheep, their use for goats is considered off-label and thereby used legally only under the direction of a veterinarian who has first-hand knowledge of the client and the goat. One dewormer approved for use in goats is Safegard/Panacur (fenbendazole). Withdrawal time for slaughter is 14 days, and withdrawal time for milking is 4 days.

Meet the Enemy

Two worm species cause major headaches for goat producers, the barber-pole worm and its cousin the brown stomach worm.

Barber-Pole Worm (Haemonchus contortus)

The barber-pole worm (also called the twisted wire worm) is a member of the gastrointestinal *Trichostrongylus* species, a group of nematodes that livestock producers often call roundworms. Its life cycle is typical of most nematode species.

1. Female worms living in a goat's abomasum release eggs, which become encapsulated in goat droppings, then fall to the ground. In one or two days the eggs hatch and, still encased in goat droppings, begin their lives as first-stage larvae. They soon molt into second-stage larvae. Both stages feed on bacteria eliminated in the droppings, while storing energy they'll need to survive later on.

2. They molt again, becoming third-stage larvae, and then, providing the dropping stays moist, they emerge and start looking for a host. If the droppings

are too hard, the third-stage larvae may become arrested (a phenomenon called *hypobiosis*), meaning they'll wait for one to three months until moisture softens the droppings and they can survive in the outside world.

3. As third-stage larvae still encased in their second-stage skin and feeding on stored energy, they climb onto growing grass and wait to be accidentally eaten by a goat or a sheep. Exactly how long they can survive depends on air temperature and the amount of energy they've stored; 30 to 90 days is the norm.

4. When a goat happens along and consumes the blade of grass to which the third-stage larvae are clinging, the larvae set up house in the animal's abomasum and molt into fourth-stage larvae. Then, if conditions are favorable (based on a number of variables including air temperature, greening of grass, rain following a drought, and estrogen spurts of their host at kidding), fourth-stage larvae molt into adults. If conditions aren't favorable, they remain arrested as fourth-stage larvae and wait for conditions to improve.

5. Two to three weeks after entering their host as third-stage larvae, adults copulate and females lay eggs, and the three-week cycle plays out again.

Barber-pole worms feed on blood. When a host becomes badly infested, it faces serious, life-threatening anemia. One thousand barber-pole worms can rob their host of up to a pint and a half of blood per day; an infestation of roughly 10,000 worms can kill a sheep or a goat. Even if they don't kill their host, heavy barber-pole worm infestations do irreparable internal damage to sheep and goats' stomachs, resulting in poor feed conversion for the rest of their lives.

BARBER-POLE WORM FACTS

- White ovaries that twist around *H. contortus*'s red, blood-filled gut make it resemble an old-time barber pole, earning the nematode its common name.
- Barber-pole worms are big enough to be seen with the naked eye: ¾ inch long and about as big around as a paper clip wire.
- A single female barber-pole worm can lay between 5,000 and 10,000 eggs *each day*.

These worms favor hot, humid conditions and they proliferate in tropical and subtropical parts of the world, including our own southeastern states. Because arrested larvae can overwinter in the abomasums of their hosts, however, barber-pole worms can be a serious summertime threat farther north.

Barber-pole worms are the goat world's public enemy number one.

Brown Stomach Worm (Teladorsagia circumcincta, formerly called Ostertagia circumcincta)

Brown stomach worm, another member of the gastrointestinal *Trichostrongylus* species, also lives in its host's abomasum, where it feeds on nutrients harvested from the stomach's mucous lining. This permanently damages the organ, which in turn affects its ability to digest nutrients, so infested goats fail to thrive. Goats can die (though they rarely do) from severe brown stomach worm infestations.

T. circumcincta is a cool-season parasite that easily overwinters in the pastures of the northern United States. Hot, dry conditions can kill larvae in the pasture, but in autumn new infestations occur from eggs laid by adults emerging from arrested fourth-stage larvae.

GASTROINTESTINAL NEMATODES

Common Name	Genus and Species	Site	Clinical Signs
Barber-pole worm and twisted wire worm	*Haemonchus contortus*	Abomasum	Amemia, bottle jaw, weakness, sudden death
Brown stomach worm	*Teladorsagia circumcincta*	Abomasum	Weight loss, diarrhea, failure to thrive
Bankrupt worm and hair worm	*Trichostrongylus*	Abomasum and small intestine	Weight loss, black (bloody) scours, failure to thrive, occasional death
Common threadworm	*Strongyloides papillosus*	Small intestine	Weight loss, diarrhea
Threadneck worm	*Nematodirus* spp.	Small intestine	Weight loss, diarrhea
Hookworm	*Bunostomum phlebotomum*	Small intestine	Weight loss, diarrhea, failure to thrive
Nodule worm	*Oesophagostomum*	Large intestine	Damage to lining of intestine

Other Gastrointestinal Nematodes

Though infestations of other gastrointestinal nematodes aren't generally as serious as infestations of "the big two," they still adversely affect the health of their hosts. Infestations involving more than one worm species (especially in conjunction with barber-pole worm or brown stomach worm) are especially troublesome — and sadly, that's the norm.

Reducing Drug Resistance

Unfortunately, there is no easy fix. With no new dewormers on the horizon, we must somehow prevent further anthelmintics resistance but also nip worm infestation in the bud. Here are some strategies that can help.

- **Avoid across-the-board deworming.** It's estimated that 20 to 30 percent of goats in most herds carry most of the worms and shed the most eggs. These goats require frequent deworming; others don't.
- **Check your goats often.** Watch for poor hair coats, bottle jaw, diarrhea, and suspicious weight loss. Every few weeks, gently pull a lower eyelid away from each goat's face and examine its mucous membrane. It should be red to dark pink; pale pink to white membranes indicate the goat is anemic and needs to be dewormed right away. Checking eyelids in this manner is the layperson's version of FAMACHA (see box on page 175).
- **Look at the eyes.** Carefully check the eye mucous membranes about one week after it rains following a period of drought. Arrested larvae bloom under these conditions.
- **Make sure your deworming program is working.** Learn to conduct your own fecal egg checks (see The Biology of the Goat in Resources) or collect droppings from goats you suspect are wormy, place each goat's offering in its own labeled plastic bag, and take these samples to the vet (only fresh-from-the-goat droppings will do, preferably pellets that haven't touched the ground). He'll prepare a solution from each sample, examine it under a microscope, and count the number of worm eggs on each slide. Ten days after deworming each goat, collect more samples and run new tests. There should be a 95 percent reduction in egg number per gram of feces. If not, drug resistance is brewing on your farm.
- **Don't rotate dewormers every time you dose your goats.** If an anthelmintic is working, use it for at least a year, or until it loses its effectiveness.
- **Weigh your goats and dose accordingly.** Never underdose! If you can't actually weigh your goats, measure them around the heart girth and refer

to our handy tape-weighing chart. (See How Much Does It Weigh, on page 177, for a table of measurement-to-pounds information.)

- **Always check with your veterinarian before using products labeled for other species.** Goats aren't dosed with the same amount of product per pound as sheep, horses, or cattle. *Always* use the correct dosage per product per goat.
- **Most goat dewormers should be delivered by mouth** (though some work effectively as injectables — call your veterinarian for a final determination). For those administered by mouth, use a dose syringe to deposit them on the back of the goat's tongue and then elevate its chin until it swallows.
- **Don't introduce drug-resistant worms to your farm.** Quarantine and deworm all incoming goats using two products from different drug classes, according to your vet's or goat mentor's recommendations. Allow these goats time to shed super-worm eggs by keeping them indoors or in a small pen away from any pasture for a minimum of 48 hours.
- **Dry-lot and fast goats in your regular herd for 12 to 24 hours prior to deworming.** This will increase drug effectiveness (water, however, should always be provided), and confine them (with feed) an additional 48 hours post deworming to prevent them from shedding worm eggs on pasture.
- **Deworm all does immediately after kidding.** Changing estrogen levels cause arrested larvae to molt and proliferate.
- **Allow your goats to browse.** Except in early morning, when plants are covered with dew, worm larvae tend to stay within three inches of the ground.
- **Rotate your pastures.** Move goats to new pasture before grass is grazed to three inches in height. Don't regraze used pastures for 90 days in the summer or 180 days in the fall and winter unless you take a cutting of hay from it (this dries out larvae) or you also graze the pasture with a close-cropping species such as horses or cattle.
- **Feed well.** Well-fed goats are more parasite-resistant than goats that aren't fed well.
- **Discuss parasite control strategies with your goat-savvy vet or county extension agent.** Set up a first-class worm-control program tailor-made for your goats' specific needs.

FAMACHA — WHAT'S *THAT*?

The FAMACHA system is an easy-to-use parasite-control program developed in South Africa for identifying *Haemonchus contortus* in small ruminants.

Rather than deworming an entire herd, FAMACHA allows producers to separate nematode-infested goats for treatment or culling. Other goats are dewormed less frequently, which slows the development of anthelmintic resistance. It also saves the producer considerable money in dewormers, fecal exams, and lost goats.

The FAMACHA system utilizes a laminated eye chart showing varying degrees of anemia as indicated by the color of the lower inner eyelid. To obtain a card, producers must attend a FAMACHA training session. For more information about FAMACHA (and other means of alternate nematode control), visit the Southern Consortium for Small Ruminant Parasite Control website (see Resources).

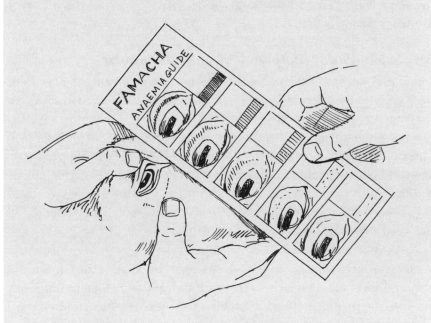

To determine a goat's level of anemia, compare the color of the goat's lower inner eyelid to those on the FAMACHA chart.

Those Other Worms

Other internal parasites that infect goats include coccidia (refer to chapter 10), tapeworm, liver fluke, and meningeal worm.

Tapeworm (Moniezia sp.)

If you see living tapeworm segments (resembling grains of white rice) in recently deposited goat droppings, don't be unduly alarmed. A major infestation may slow growth in kids, but unless an intestinal blockage occurs, tapeworms aren't a primary problem for goats.

Liver Fluke (Fasciola hepatica)

Liver flukes pose a serious problem in the perennial wetlands of the Southeast and occasionally farther north as well. They can live in the host up to ten years, feeding off the bile duct lining, and over time can cause enough irritation to lead to scarring and cirrhosis of the liver and, possibly, death. Symptoms of liver fluke infestation include anemia, severe weight loss, and low milk yields. To reach infection stage, liver flukes must be carried by snails as the intermediate hosts.

Meningeal Worm (Parelaphostrongylus tenuis)

Meningeal worm is sometimes called deer worm or meningeal deer worm because its natural host is the white-tailed deer. Goats (as well as sheep, llamas, alpacas, and wild species such as moose) are at risk wherever white-tail deer are present. Tiny, ground-dwelling slugs and snails act as the intermediate host between deer and other ruminant species. While the worm doesn't seem to bother its natural host very much, in other species, larvae migrate throughout to the host's spinal cord and brain, causing rear-leg weakness, staggering gait, hypermetria (exaggerated stepping motions), circling, gradual weight loss, and paralysis leading to death.

In areas where the parasite is endemic, many vets prescribe off-label preventive Ivermectin injections at 30-day intervals throughout the transmission season (spring and summer). Treatment is difficult and generally unsuccessful. If you live in proximity to white-tailed deer, discuss meningeal worm with your vet; its incidence in goats is on the rise.

HOW MUCH DOES IT WEIGH?

Tape-weigh your goats with this surprisingly accurate chart, developed in Australia by Boer goat breeders. Pass a cloth seamstress tape around a goat's heart girth, directly behind its front legs. Round down the resulting measurement to the next full number and compare the measurement with the chart — it's that easy!

Inches	Pounds
15	9.5
16	11.0
17	13.0
18	15.5
19	18.0
20	20.5
21	24.0
22	28.5
23	33.5
24	37.5
25	43.0
26	47.5
27	54.0
28	59.5
29	66.5
30	74.0
31	80.0
32	88.5
33	94.5
34	100.0
35	109.0
36	117.0
37	123.5
38	131.0
39	136.5
40	141.0
41	148.0

External Parasites

External parasites such as flies, ticks, and lice feed on body tissue (blood, skin, hair) and can transmit diseases from sick to healthy animals. External parasites can reduce weight gains and milk production while making life unpleasant for their caprine hosts.

Nose Bots

The nose bot fly lays its first-stage larvae in the nostrils of sheep and goats. These migrate up the nasal passages and feed on mucus and mucous membranes. Infested goats lose their appetite, shake their head, and grind their teeth, and there is usually an opaque discharge from infested nostrils. In the presence of the adult fly, goats are excited; they shake their heads, snort, and rush around with their nose in the dust. Currently, Ivermectin drench is the only effective treatment for nose bots.

Did You Know?

- The nose bot is a hairy, yellowish fly about the size of a common housefly. Nose bot flies are often mistaken for bees.
- A single female bot fly may deposit as many as 500 larvae in the nostrils of unwary sheep and goats.
- Bot fly larvae remain in their hosts' sinuses for eight to ten months and then are sneezed out of the nostrils. Think about *that* the next time you hear a goat sneeze!

Lice

Lice spend their entire life on their host. Both immature and adult stages suck blood and feed on skin. Goat lice are host-specific and infest only goats and sheep.

Louse-infested animals have dull, matted coats and display excessive scratching and grooming behavior, resulting in raw patches on the skin and loss of hair. Weight loss may occur when infested animals don't eat because of reduced appetite and incessant scratching. Heavy infestation can reduce milk production by 25 percent. A louse-infested goat is often listless, and in severe cases the loss of blood to sucking lice can lead to serious anemia.

There are two types of lice: sucking lice (these pierce the host's skin and draw blood) and biting lice (these have chewing mouthparts and feed on particles of hair and scabs). Lice are generally spread via direct contact, often when infested animals join an existing herd. Populations vary seasonally. Most sucking and biting lice proliferate during autumn and reach peak numbers in late winter or early spring. Summer infestations are rare.

Three varieties of sucking lice infest goats:

- African blue lice are a problem only in semitropical parts of the United States. They're usually found on the goat's body, head, and neck. Heavy populations can kill a goat.
- Foot lice infest the feet and legs of goats and sheep. When populations peak in the spring, lice may infest the belly area too, and scrotum infestations on bucks are relatively common. Kids, however, tend to have the highest rate of infestations.
- Goat sucking lice occur all over the animal's body. Sheep can get them as well.

Biting lice eggs hatch in 9 to 12 days, with an entire life cycle taking just one month. Wintertime infestations are usually the most severe.

Treatment, generally with an over-the-counter residual pesticide designed for livestock, is needed whenever an animal scratches and rubs to excess. Louse control is difficult because pesticides kill lice but not their eggs. Given that eggs of most species hatch 8 to 12 days after pesticide application, retreatment is necessary 2 or 3 weeks following the first treatment.

Mites

Itch or mange mites feed on the skin surface or burrow into it, making minute, winding tunnels from one-tenth to one inch long. Fluid discharged at the mouth of each tunnel dries and forms a scab. Mites also secrete a toxin that causes intense itching. Infested goats rub and scratch themselves raw. Infestations are highly contagious; if one goat has mites, treat the whole herd.

12

Livestock Guardians

GOAT PRODUCERS EVERYWHERE AGREE: It's a heartbreaking, financial disaster when predators raid your flock and kill your goats. And it will happen to you if you don't add livestock guardians to your herd.

Statistics Tell the Tale

The USDA's National Agriculture Statistics Service (NASS) keeps track of what America's sheep and goats die from and periodically publishes findings in a report titled "Sheep and Goats Death Loss." According to the edition issued on May 6, 2005, predators killed 155,000 goats during 2004, accounting for slightly more than 37 percent of goat deaths that year. Sheep figures are just as astounding: 225,000 sheep were killed by predators during 2004 (also accounting for 37 percent of total losses).

Of the goats killed, 40,000 were adults. Small ruminants (sheep and goats combined) were killed mainly by coyotes (more than 60 percent of the total) but also by dogs, mountain lions, bears, foxes, eagles, bobcats, and other species (among them wolves, ravens, and black vultures).

Predation is a serious problem and one you'll have to address in order to raise goats anywhere in North America — even in relatively populated areas, where free-roaming dog predation may pose a major risk.

WHAT'S KILLING MY GOATS?

Until you put effective livestock guardians in place (and possibly, to a lesser degree, even after), you're likely to lose goats to roving predators. It's important to identify the culprits involved. The behaviors outlined below are typical of North America's most common goat predators.

BLACK BEARS

- Kill with crushing bites to the side of the neck, spine, and throat.
- Frequently kill more than one goat.
- Leave claw marks on the neck, back, and shoulders.
- Consume the udder and flank region first.

BOBCATS

- Usually kill small kids by biting the backs of their necks; they're more likely to leap on the backs and bite the necks and throats of larger kids.
- Begin feeding on internal organs after opening the body cavity behind the ribs.
- Often drag and cover their kill.

COUGARS

- Generally bite into the back of the neck and skull, causing massive hemorrhaging; broken necks are common in cougar kills.
- Leave large claw marks on the head, neck, shoulders, and flanks.
- Usually gut the goat and drag the entrails aside, then consume the heart, lungs, liver, and larger leg muscles.
- Sometimes drag and cover their kill.

COYOTES

- Typically bite the adult goat's throat just behind the jaw and below the ear, but they may kill kids by biting the head, neck, or back, causing massive tissue and bone damage.
- Generally begin feeding on the flank just behind the ribs, first consuming internal organs.
- Often consume the animal's nose, especially when feeding on a small kid.

continued on next page

WHAT'S KILLING MY GOATS? *(continued)*

DOGS

- Indiscriminately mutilate multiple prey (often without killing it), leaving deep bites on multiple body parts.
- Rarely feed on the animals they kill.

EAGLES

- Leave talon punctures on the head and body; opposing talon punctures are located four to six inches from the middle talon wound.
- Consume organs; sometimes split the skull and eat an animal's brain.
- "Skin out" the carcass and strew clumps of hair around the kill.
- Leave patches of streaky white droppings.

FOXES

- Usually attack the throat of kids, but some kill by multiple bites to the neck and back.

VULTURES, CROWS, RAVENS, MAGPIES

- Attack the eyes and nose, navel, and anal area of newborn kids. They typically blind their prey by pecking out its eyes, even if they don't kill it.

Protecting the Herd

When the National Agriculture Statistics Service surveyed sheep and goat producers in 2004 to ask how they managed predation, more than 50 percent indicated they relied (at least in part) on predator-proof fencing, 33 percent also penned their stock at night, and 55 percent kept livestock guardian animals with their flocks and herds.

The concept of livestock guardian animals goes back a long, long way — about 6,000 years ago — to the mountain regions of Turkey, Iraq, and Syria, where guardian dogs were first trained to protect sheep and goats. Today's farmers and ranchers still use guardian dogs to protect their livestock, but they've added guardian donkeys and llamas to the mix.

Donkeys

Donkeys are sweet and charming animals, but they're born with an ingrained hatred for the canine tribe. Therefore, some donkeys make first-rate herd guardians where coyotes and dogs are troublesome, the keyword being *some*. (See the caveats in the following bulleted list.)

Donkeys require no specialized training; almost any donkey will attack dogs and coyotes. They chase while braying, biting, and sometimes kicking at canine invaders. Donkeys have keen hearing and good eyesight, so dogs and coyotes rarely sneak past them when they are on guard.

Donkeys are hardy; long-lived (25- to 30-year life spans are the norm); and with the exception of feed medicated with Rumensin (which is poisonous to equines of all kinds), they eat the same sorts of things you probably already feed your goats.

Finally, you won't break the bank buying a guardian-quality, standard-size or larger donkey (miniatures are too small to do anything but bray an alarm). You can adopt an inexpensive, easy-to-tame wild burro from the U.S. Department of the Interior's Bureau of Land Management (BLM) National Wild Horse and Burro Program or from a donkey rescue group or buy one locally for anywhere from $150 to $800, depending on quality, registration status, and size.

But before you do, also consider the donkey disadvantages:

- Gelded (castrated male) donkeys make the best guardians, and jennies (females) run a close second best. But jacks (intact males) simply won't do; some are aggressive toward humans, most will savage goats they dislike, and many have been known to kill newborn kids. Also, some jacks try to breed the does they protect, inflicting serious, sometimes fatal, injuries in the process.
- Donkeys prefer their own kind. If you have more than one donkey, they'll generally seek out each other's company and ignore the goats. Many prefer the company of horses to goats, as well. If you choose a guardian donkey, don't plan on keeping other equines near the herd.
- Unlike guardian dogs, donkeys haven't been bred for generations to look after other livestock; some simply aren't interested in bonding with goats.

Thus, if you buy a donkey to guard your herd, ask if you can return it if it's aggressive toward or uninterested in your goats. Yet in one survey (reported in the Colorado State University publication "Livestock Guard Dogs, Llamas and Donkeys"), 59 percent of Texas producers who use guardian donkeys rated them good or fair for deterring coyote predation, and another 20 percent rated them excellent or good. In addition, 9 percent of the sheep and goat producers polled for the 2004 National Agriculture Statistics Service survey successfully keep them. In the right sort of setting, guardian donkeys can work.

Llamas

In 1990, researchers at Iowa State University polled 145 sheep producers in five western states to determine the effectiveness of llamas for reducing dog and coyote predation. Before llamas joined their flocks, they reported an average annual loss of 21 percent of their ewes and lambs, which was reduced to 7 percent after llamas were employed. Eighty percent rated their llamas as effective or very effective for guarding sheep. In another study, conducted in Utah, 90 percent of producers rated guardian llamas as effective or very effective on the job. National Agriculture Statistics Service figures indicate that 14 percent of sheep and goat producers used llamas as guardians in 2004.

The pros and cons of having llamas on guard are similar to those of using donkeys. Llamas naturally dislike dogs and coyotes, and while most make excellent guardians, some don't. Llamas prefer the company of other llamas, so you usually have to keep just one per pen or herd or they'll pal around and ignore the goats. Intact males may attempt to breed does, injuring or even smothering them in the process. But llamas have certain advantages and disadvantages that are quite different from those of donkeys:

- Llamas require the same food (it's safe to feed them Rumensin-medicated products), vaccinations, and hoof care that goats do. It's easy to treat them as just another member of the herd.
- The "wool" of long-haired llamas can be shorn and hand-spun into yarn. Use the fleece yourself or sell it.
- On the other side of the coin, llamas are more aloof than donkeys. Where most donkeys crave human interaction, the average llama doesn't. This makes it more difficult to catch and handle one for routine maintenance chores.

- Llamas don't do as well as donkeys in hot, muggy climates, where they're prone to heat exhaustion. Depending on the length of their fleece, most llamas require full or partial shearing at least once a year.
- Llamas don't live as long as donkeys. The average llama life span is 10 to 15 years.

A guardian-quality gelded llama costs $200 to $700; females usually sell for somewhat more. You may consider adopting a llama from a llama rescue organization. Nowadays some rescues test their charges with sheep before placing them in homes as guardians, and, in any case, will take them back if they don't work out. Investigate adopting a guardian llama via the Southeast Llama Rescue website.

For more information about donkeys and llamas as guardians for your goats, see Resources.

Dogs

For thousands of years stalwart European, Middle Eastern, and Asian guard dogs of dozens of types and breeds have watched over herds of goats and flocks of sheep, protecting them from predation by wolves, bears, jackals, and human thieves.

Some of these dogs eventually made their way to North America. By the mid-1980s, when wildlife biologists Jeffrey Green and Roger Woodruff, of the U.S. Sheep Experiment Station in Dubois, Idaho, conducted a survey on the effectiveness of livestock guardian dogs (LGDs), these animals were already commonplace in western flocks and herds. As part of their survey, Woodruff and Green sent questionnaires to 948 sheep and goat producers in 47 states and seven Canadian provinces. They received 399 responses regarding 763 dogs. Nonrespondents they phoned reported on 45 additional dogs. These are their findings.

- Ninety percent of respondents grazed their livestock in pasture settings and 10 percent on rangeland.
- Among small-pasture operators (those who owned between 4 and 50 head of stock), 78 grazed sheep, 26 raised goats, and 11 had both.
- Large-pasture operators (who owned 56 to 8,000 animals) totaled 175 producers who raised sheep, 22 who ran goats, and 11 who owned both.

- Among rangeland operators (who owned 1,200 to 16,000 animals), 33 had sheep and 4 had goats.

When asked about their livestock guardian dogs' effectiveness, these sheep and goat producers reported as follows.

- Ninety-nine percent of pasture operators recommended the use of guardian dogs and 1 percent did not. Seventy-one percent rated their dogs as very effective, 21 percent as somewhat effective, and 9 percent as not effective at all.
- Thirty-eight of 39 rangeland operators recommended the use of dogs; one declined to comment. Sixty-six percent rated their dogs very effective, 19 percent as somewhat effective, and 15 percent as not effective at all.

More than 95 percent of the dogs involved in this survey were recognized guardian breeds. Great Pyrenees (437 dogs) and Komondorok (138 dogs) headed the list; Akbash, Anatolian, Maremma, Shar, Kuvasz, and crosses were also represented. Results were tallied regarding each breed's effectiveness and aggressiveness and problems encountered with each.

According to current NASS figures, nearly 32 percent of American sheep and goat producers use livestock guardian dogs representing a score or more breeds. Let's look at some common guardian breeds along with a few that aren't as common.

Anatolian Shepherd

Origin: Semiarid Anatolian Plateau of central Turkey
Name in native land: Anadolu Coban Kopegi (Anatolian shepherd's dog)
Height: Dogs 28–35 inches; bitches 26–28 inches
Weight: Dogs 100–160 pounds; bitches 90–130 pounds
Coat type: Short to medium double coat
Color: Usually fawn with a black mask, but any color is acceptable
Life span: 10–15 years
Green and Woodruff survey: Fifty-six Anatolians were evaluated by respondents. Among owners, 77 percent rated Anatolians very effective guardians, 13 percent as somewhat effective, and 10 percent as not effective at all. Ninety-six percent of the dogs were assessed as aggressive toward predators and 86 percent toward dogs. Only 10 percent of owners experienced major problems

with their Anatolians, 48 percent had minor problems, and 42 percent had no problems whatsoever. Sixty-nine percent of Anatolians stayed most of the time with the stock they were supposed to guard, 16 percent usually stayed with them, and 15 percent rarely stayed to guard their charges. Nine percent of the Anatolians in the study bit people, while 14 percent injured sheep or goats.

Notes: These dogs are slow-maturing and sensitive to anesthesia. Hip dysplasia isn't as common as in other large breeds. The breed is recognized by the AKC and the UKC.

ANATOLIAN SHEPHERD

Akbash Dog

Origin: The Akbash region of Turkey
Name in native land: Akbas (white head); Coban Kopegi
(pronounced CHO-bawn CO-pay, meaning "shepherd's dog")
Height: 28–32 inches
Weight: 90–130 pounds
Coat type: Short to medium double coat
Color: White
Life span: 10 or 11 years
Green and Woodruff survey: Sixty-two were evaluated by respondents. Among owners, 69 percent rated Akbash dogs very effective guardians, 22 percent as somewhat effective, and 9 percent as not effective at all. One hundred percent of these dogs were aggressive toward predators and 92 percent toward dogs. Fifteen percent of

AKBASH DOG

owners experienced major problems with their Akbash dogs, 49 percent had minor problems, and 36 percent had no problems whatsoever. Seventy-one percent of Akbash dogs stayed most of the time with the stock they were supposed to guard, 12 percent usually stayed with them, and 17 percent rarely stayed to guard their charges. Six percent of the Akbash dogs in the study bit people, while 20 percent injured sheep or goats.

Notes: Hip dysplasia isn't as common as in other large breeds. The breed is recognized by the UKC.

Caucasian Ovcharka (Caucasian Mountain Dog; Rare Breed)

Origin: The Caucasus Mountain regions of Armenia, Azerbaijan, Georgia, and Russia

Name in native land: Kavkazskaya ovcharka in Russian, Nagazi in the Georgian Republic, and Gampr in Armenia

Height: 25–32 inches

Weight: 100–150 pounds

Coat type: Short or long double-coated; abundant ruff and fringing

Color: Usually agouti gray; otherwise, any color except red and white like the Saint Bernard, solid black or brown, or solid black and tan

Life span: 10 or 11 years

Green and Woodruff survey: Not listed

Notes: *Ovcharka* (also spelled ovtcharka and owtcharka and pronounced uhf-CHAR-k) is a Russian word meaning "shepherd's dog." These dogs are slow-maturing and somewhat prone to hip and elbow dysplasia. Otherwise vigorously healthy, the Caucasian Ovcharka was extensively used as a military guard dog throughout the former Soviet Union.

CAUCASIAN OVCHARKA

Great Pyrenees (Pyrenean Mountain Dog)

Origin: The Pyrenees Mountain region of southern France
Name in native land: Chien des Pyrénées, Chien de Montagne des Pyrénées
American nickname: Pyr
Height: Dogs 27–32 inches; bitches 25–32 inches
Weight: Dogs from 100 pounds; bitches from 80 pounds
Coat type: Long, coarse outer coat may be straight or slightly wavy; fine, soft, and thick undercoat
Color: White, often having gray or tan "badger" markings on the head
Life span: 10 or 11 years
Green and Woodruff survey: Four hundred and thirty-seven were evaluated by respondents. Among owners, 71 percent rated Great Pyrenees as very effective guardians, 22 percent as somewhat effective, and 7 percent as not effective at all. Ninety-five percent of these dogs were aggressive toward predators and 67 percent toward dogs. Eleven percent of owners experienced major problems with their Great Pyrenees, 47 percent had minor problems, and 42 percent had no problems whatsoever. Fifty-three percent of Great Pyrenees stayed most of the time with the stock they were supposed to guard, 24 percent usually stayed with them, and 23 percent rarely stayed to guard their charges. Four percent of the Great Pyrenees in the study bit people, while 7 percent injured sheep or goats.

Notes: Pyrs have double dewclaws on their hind legs. They are somewhat prone to hip dysplasia and occasionally to skin problems. The breed is recognized by the AKC and the UKC.

GREAT PYRENEES

Kangal Dog (Rare Breed)

Origin: Kangal District of Sivas Province in central Turkey
Name in native land: Karabash (black head)
Height: Dogs 30–32 inches; bitches 28–30 inches
Weight: Dogs 110–145 pounds; bitches 90–120 pounds
Coat type: Short, dense double coat

Color: Light dun to gray with black mask and ears
Life span: 11–14 years
Green and Woodruff survey: Not listed

Notes: While some authorities lump together Kangal dogs and Anatolian shepherds, though they are similar breeds, each is distinct. According to the Turkish Dogs website FAQ:

KANGAL DOG

- The lineal and weight requirements for Kangal dogs and Anatolians are different.
- Kangal dog color is restricted, while Anatolians are acceptable in all colors and markings.
- The Kangal dog is short-coated while the Anatolian coat may be up to four inches in length.
- Kangal dogs registered with the UKC have originated or descended from dogs obtained from the Sivas-Kangal region of Turkey. Anatolians appear to have originated in any area of Turkey.
- The temperaments of the two dogs are different: The Kangal dog is people-oriented and hostile only to traditional predators.
- The Kangal dog breed is recognized by the UKC.

Karakachan (Rare Breed)

Origin: Bulgaria
Name in native land: Karakachansko Kuche
Height: Dogs 25–29 inches; bitches 23½–27½ inches
Weight: Dogs 85–110 pounds, bitches 75–100 pounds
Coat type: Short (hair up to about 2⅓ inches in length); long (anything over 2⅓ inches); ruff, feathering on legs, long hair on tail; double-coated

KARAKACHAN

Color: Most are spotted, but solid-colored dogs are acceptable; colored areas can be black, gray, brown, yellow, or brindle.

Life span: 12–16 years

Green and Woodruff survey: Not listed

Notes: This breed is critically endangered worldwide; breeders in the United States are working to preserve this extremely ancient, effective livestock guardian breed before it becomes extinct.

Komondor (Hungarian Sheepdog)

Origin: Hungary

Name in native land: Komondor (pronounced KOM-on-door); plural is Komondorok; may be derived from the term *komondor kedvu,* meaning "somber" or "angry"

Height: 25½ inches and up

Weight: Dogs up to 125 pounds; bitches 10 percent less

Coat type: Very long and nonshedding double coat, divided into sections and allowed to form felted cords for show; working Komondorok are usually clipped once a year

Color: White

Life span: 10–12 years

Green and Woodruff survey: One hundred and thirty-eight were evaluated by respondents. Among owners, 69 percent rated Komondorok as very effective guardians; 1 percent as somewhat effective; and 12 percent as not effective at all. Ninety-four percent were aggressive toward predators and 77 percent toward dogs. Fourteen percent of owners experienced major problems with their Komondorok; 48 percent had minor problems; and 38 percent had no problems whatsoever. Fifty percent of Komondorok stayed most of the time with the stock they were supposed to guard; 23 percent usually stayed with them; and 27 percent rarely stayed to guard their charges.

KOMONDOR

Seventeen percent of the Komondorok in the study bit people, while 24 percent injured sheep or goats.

Notes: Komondorok are prone to hip dysplasia, bloat, and skin diseases. They are one of the strongest and most aggressive of the livestock guardian breeds — and are also one of the most effective in guarding herds against human thieves. The breed is recognized by the AKC and the UKC.

Kuvasz

Origin: Hungary
Name in native land: Kuvasz (pronounced KOO-vahz); plural is Kuvaszok
Height: Dogs 28–30 inches; bitches 26–28 inches
Weight: Dogs 100–115 pounds; bitches 70–90 pounds
Coat type: Medium, straight or wavy; thick undercoat
Color: White (over dark skin)
Life span: 10–12 years
Green and Woodruff survey: Only seven Kuvaszok were evaluated by respondents. Among owners, 57 percent rated Kuvaszok as very effective guardians; 29 percent as somewhat effective; and 14 percent as not effective at all. One hundred percent were aggressive toward predators and 67 percent toward dogs. Fourteen percent of owners experienced major problems with their Kuvaszok and 86 percent had minor problems. Thirty-three percent of Kuvaszok stayed most of the time with the stock they were supposed to guard, 33 percent usually stayed with them, and 34 percent rarely stayed to guard their charges. None of the Kuvaszok in the study bit people, while 40 percent injured sheep or goats.

Notes: Kuvaszok are prone to hip dysplasia, several other minor lameness issues, skin problems, and allergic reactions, and they may drool. The breed is recognized by the AKC and the UKC.

KUVASZ

Maremma Sheepdog

Origin: Italy

Name in native land: Maremma (pronounced mair-EM-ma), Abruzzese, Cane da Pastore (sheepdog), Mareemmano-Abruzzese

Height: 23½–28½ inches

Weight: 66–100 pounds

Coat type: Long, harsh (versus soft) double coat; outer coat is straight or slightly wavy

Color: White with ivory, light yellow, or pale orange on the ears

Life span: 10–12 years

Green and Woodruff survey: Twenty Maremma sheepdogs were evaluated by respondents. Among owners, 70 percent rated Maremma sheepdogs very effective guardians; 20 percent as somewhat effective; ten percent as not effective at all. Ninety-four percent were aggressive toward predators and 94 percent toward dogs. Eighteen percent of owners experienced major problems with their Maremma sheepdogs, 24 percent had minor problems, and 58 percent had no problems at all. Seventy-nine percent of Maremma sheepdogs stayed most of the time with the stock they were supposed to guard, 16 percent usually stayed with them, and 5 percent rarely stayed to guard their charges. Five percent of Maremma sheepdogs in the study bit people, while 20 percent injured sheep or goats.

Note: The breed is recognized by the AKC and the UKC.

MAREMMA SHEEPDOG

Polish Tatra Sheepdog

Origin: The Podhale section of the Tatra region of the Carpathian Mountains in the south of Poland

Name in native land: Polski Owczarek Podhalanski

Height: Dogs 26–29 inches; bitches 24–26

Weight: 80–130 pounds

Coat type: Heavy double coat, with a top coat that is hard to the touch, straight, or slightly wavy; profuse, dense undercoat

Color: White over dark skin

Life span: 10–12 years

Green and Woodruff survey: Not listed

POLISH TATRA SHEEPDOG

Notes: Occasional hip or elbow dysplasia, bloat, skin problems, occasional epilepsy or cataracts. The breed is recognized by the UKC.

Whichever breed (or combination of guardian dog breeds) you choose, here are a few points to keep in mind.

All dogs of the livestock guardian breeds are extremely large, intelligent, strong-willed, and potentially aggressive and dominant — they aren't to be handled casually by the faint of heart. Most are dog-aggressive; they sometimes kill household pets that wander in with the goats, and some fight to the death with other guardian dogs of the same sex. Because most predators strike at night, livestock guardian dogs are often active after dark. Barking is their primary means of warning off intruders, so expect a lot of night barking from your LGDs. Before buying a livestock guardian dog, visit LGD breeders and sheep or goat producers who keep them as flock and herd guardians. Compare philosophies and training methods. Know what you're getting into before you bring home a dog.

To be effective, livestock guardian dogs must be bonded to and stay with the livestock they're expected to guard. During a puppy's prime socialization period (from four weeks to 14 weeks of age), minimize human interaction with LGD puppies — but don't believe sheep and goat producers who insist you should never play with or cuddle a guardian puppy. To handle them for routine deworming, nail trimming, and so forth as adults, LGD puppies *must* look forward to interacting with humans. It's important, however, to play with a puppy on its own turf (the goat yard), rather than taking it to the house or for rides in the truck; it needs to understand that watching the goats is its primary job.

Livestock guardian puppies needn't be trained to watch over sheep and goats; the desire to guard has been bred into these animals for thousands of years. Most begin showing guardian dog behaviors by five or six months of age (scent marking, purposeful barking, increased interest in the goats, deliberate patrolling), although few become reliable protectors until they achieve mental and physical maturity, usually at about two years of age.

As they pass through adolescence, most LGD puppies exhibit inappropriate behavior toward their goats, usually in the form of play chasing. Expect it and work through this phase; there's an effective working livestock guardian waiting on the other side.

Occasionally, a pup comes along that has no interest in guarding goats — or worse, doesn't like them. All breed organizations maintain rescue programs for their dogs; please place nonworking livestock guardian dogs with appropriate groups. Don't put them down: They often make excellent pets.

Producers Profile

Dave and Judy Muska, Lazy J Goat Farm
Schulenburg, Texas

Though Judy and Dave Muska no longer raise exquisite Boer goats at their Lazy J Goat Farm in south-central Texas, Judy still participates at most of the major goat-oriented e-mail lists, where she's one of the first to respond to a fellow goat owner in need.

Q *Judy, how did you and Dave get your start in goats?*

A A little over fifteen years ago, a neighbor down the road called me and said she had a little newborn goat whose mother had died. She said she couldn't care for her and asked if I would take her. Never having seen a goat in person, I went and got her. That's how my love affair with goats began.

Q *Can you tell us a little about your breeding program?*

A We started slowly and have worked our way to a herd of about 70 goats. I know we made every mistake that was possible to make, but through those mistakes, I've learned about goats and their proper care. We've culled heavily and now have a herd we're very proud of. We have fullblood Boer goats and LaMancha dairy goats to supplement our babies. We have a CL- and CAE-free herd and follow a strict vaccination and deworming program. Our goats show well, too. In the past two years they've won three Grand Champions, two Reserve Grand Champions, and over 15 first places. In addition to our goats, we breed, sell, and use registered Anatolian shepherd and Great Pyrenees livestock guardian dogs, and we also have a special love for our Chinese pugs, which we breed, raise, and sell.

Q *And how did you and Dave decide to raise Boer goats?*

A We had already been raising goats about five years before the Boer came to the United States. In 1993 Boer bucks were selling for $30,000 and up. A neighbor not far from us got a Boer buck during the latter part of 1993, so we went down to see him. He had not only the buck, but a doe with twins about three weeks old as well. It was love at first sight for me. I had never seen goats near that big or that beautiful — and those babies were so tame! I knew right then that somehow, some way I was going to have them.

Q *What's your take on today's Boer goat industry?*

A I think the goat industry is going strong. Yes, prices have dropped, but interest in Boers has soared. More people are getting into Boer goats every day. Most meat packers won't even look at something without a red head and a white body. Boers are here to stay, and they've already made a huge impact in the goat world. No other goat has even come close.

Q *What would be your best advice for newcomers to the Boer goat world?*

A The best advice I could give any newcomer is to find out about the people you're thinking of buying from. Ask for references. Talk with people who've bought from them. If they aren't hiding anything, they should have no problem with this. If you're buying breeding stock and paying a considerable amount, ask the seller if he or she will guarantee in writing that the animals are CL- and CAE-negative. That seller owes it to you to be selling you a healthy animal.

Q *Judy, since you raise livestock guardian dogs, could you share a few words about them with our readers?*

A I've been raising these wonderful guardians for almost 15 years. My breeding stock came from long working bloodlines. I don't believe in that myth of not naming or handling a pup; in other words, some people say "Don't make friends with it." That's false information! I handle all my pups from birth. They're born and raised with the goats, and they get attention from me every day, but right out there with the goats. I never bring them in the house, because then they know their place is with the goats, and then that's where they want to be.

13

Breeding
Meat Goats

WHETHER OR NOT YOU SUCCEED as a goat rancher will depend on how well you manage the breeding aspect of your goat business. In the "olden days," a billy was released with a herd of nannies and five months later there were kids. Nowadays, in a world where bucks may cost thousands of dollars and does must deliver and raise healthy twins or better for producers to show a decent profit, things are considerably more involved.

Bucks Are Half the Herd

Each doe in a herd influences one to four or so offspring per year, but the buck puts his stamp on every kid. If you want to make money raising goats, good bucks will certainly help you do it. Buy the best bucks you can afford, based on what you need them to do. A brawny crossbred buck that consistently sires fast-gaining, meaty kids works better in a commercial situation than a modern, show-quality Boer wether sire. Both of these bucks, however, should be healthy, well (but not overly) fed, and managed to work at prime capacity.

Prior to the beginning of rut, each buck should be given an impromptu health exam. Does he appear to be healthy (refer to the Healthy Goat, Sick Goat chart in chapter 5)? Palpate his testicles, checking for size, symmetry, and texture. The larger his testicles, the better his sperm production will be.

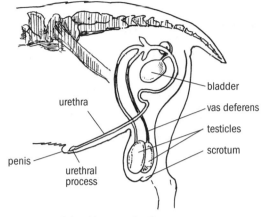

A buck's reproductive system

Both testicles should be the same size and resilient to the touch and should feel as firm and warm as muscle tissue.

Check his *prepuce* (the end of the penis sheath) and penis for infection or injury that might make it too painful for him to breed. For instance, bucks on a high-protein diet, especially bucks that are fed early-cutting, leafy alfalfa hay, are prone to a condition called pizzle rot *(posthitis)*. Their urine is high in urea and thus very alkaline, which scalds the prepuce. In addition, hair in the inflamed area traps moisture, allowing *Corynebacterium* spp. bacteria to invade and proliferate. The prepuce becomes ulcerated, swollen, and very painful. This buck usually won't breed does.

LEADING CAUSES OF INFERTILITY IN BUCKS

- Disease or injury
- Physical defects (congenital or acquired)
- Too fat or too thin
- Improper feeding
- Poor libido
- Stress (including heat stress and the temporary aftereffects of high fever)
- Age (too young or too old)

DO-IT-YOURSELF SEMEN EVALUATION

While it's not as conclusive as having sperm evaluated by a vet or an artificial insemination collector, visually examining your buck's semen can give you an indication of its quality. High-quality semen is thick and milky; marginal semen is thin and milky; and low-quality semen is clear- to amber-colored.

A buck in A1 health and in the prime of his life can impregnate 100 or more does in a single breeding season, but it's best not to overwork your bucks. A hardworking buck may forget about eating to the detriment of his health, and if he hasn't time to recuperate between breeding sessions, he'll be less fertile than the norm. Remember, after just three ejaculations a buck's sperm count is reduced, and after six to eight ejaculations, he'll be shooting immature sperm incapable of fertilizing any does.

Bucks should be handled carefully during breeding season; rut-induced aggression sometimes occurs. If a buck is testy outside of breeding season, *never* turn your back on him while his hormones are raging. A well-known producer tells how an enraged Boer buck chased him to his truck, then slammed into the closed door hard enough to cave it in and heave the truck four feet sideways. These are immensely powerful, testosterone-driven animals with long, lethal horns on their heads. Be careful. A single miscalculation can put you in the hospital — or worse.

If your buck is a sweet-natured, easygoing fellow, watch him anyway, because he won't be himself during rut. Maniac or sweetheart, don't allow *any* buck to move between you and your only avenue of escape, whether in pen, paddock, or pasture. If possible, feed from the outside of enclosures that hold bucks in rut. If you must go in, carry something with which you can defend yourself should the need arise. While some producers carry weapons such as a shock prod and pepper spray (and sometimes are forced to use them), a blast-to-the-face stream of liquid from a spray bottle filled with vinegar-laced water will divert most bucks.

Because some bucks consider their caretakers a part of the herd, women and girls should be especially wary of bucks in rut. A big, smelly, amorous buck looming over you and blubbering in your face is definitely a situation to avoid.

If you find yourself in a situation you'd like to escape intact, grab that buck's beard and hang on tight. Walk him (he won't argue about it) to your nearest means of escape and don't let go until you're safe. *Hint:* Afterward, wash your hands with Fast Orange hand cleaner from the auto parts store — it's the only readily available product I know of that efficiently removes buck stench from human skin.

In an emergency, grab an aggressive buck's beard and hang on!

BREAKING FREE

When a baby buckling is born, his *urethral process* and the *glans* of his penis are attached to the inside of his prepuce (sheath) by the *frenulum membrane.* As his body begins producing testosterone (sometimes when he's only a few weeks old), he'll begin "practice-breeding" his dam and siblings and may engage in a bizarre-looking form of air-humping behavior in which he semi-squats and then repeatedly thrusts his hips. These sessions help break down the adhesion and allow him to extend his penis. Once this happens, he's probably capable of impregnating females. For this reason, many goat producers wean bucklings when they're only eight weeks old (doelings are generally left on their dams until they're 10 to 12 weeks of age).

Urinary Calculi

Urinary calculi (also called uroliths and bladder or kidney stones; the disease they cause is called *urolithiasis*, or UC) are hard masses of mineral crystals that form in the urinary tract of sheep and goats. Does generate urinary calculi, but because of a female's relatively shorter, straighter urethra (the tube that empties urine from the bladder), she can pass these stones with little or no discomfort. Males aren't as lucky. Stones easily lodge in a male's longer, slenderer urethra, especially where the small-diameter urethral process (colloquially known as the pizzle) extends beyond the penis.

When this happens, the animal can no longer urinate (in partial blockages, a limited flow is maintained), and unless the condition is quickly corrected, uremic poisoning sets in, the bladder or urethra ruptures, and the buck dies. A goat suffering from a blockage caused by urinary calculi will display at least some of these symptoms:

- Difficult, painful, and/or dribbling urination
- Blood in urine
- Straining, abdominal contractions, tail twitching, kicking at the abdomen
- Crying out or groaning in pain
- Standing in a rocking-horse stance, with front legs perpendicular to the ground and hind legs angled out in back
- Crystals in the hairs around the prepuce
- Swollen or distended penis; palpation makes the goat cry out in pain
- Lack of appetite, depression, collapse, and ultimately death

Most but certainly not all bucks and wethers that develop urinary calculi are on high-grain, low-forage diets or diets in which legume forage, especially alfalfa, plays a primary role. Goat rations should be formulated in a way that they provide a 2:1 calcium-to-phosphorus ratio (two parts calcium to one part phosphorus); otherwise, males are apt to develop phosphatic calculi, the most commonly encountered type of urinary stones. It's important to feed bucks and wethers a ration based on browse (when available), good grass hay or similar forage, and a quality mineral supplement designed to balance the mix.

In addition, water intake is essential for a healthy, dilute urine flow. When goats don't drink enough, urine becomes overly concentrated and crystals start to form. Male goats should always have free access to a fresh, palatable water supply kept reasonably cool in the summer, liquid in the winter, and inside the shelter during periods of inclement weather (goats won't troop to the outside water trough if it means they have to get wet). Again, if you wouldn't drink from a buck's water source, chances are he won't either.

Other strategies to reduce the instance of urinary calculi include:

- Adding loose salt to a goat's diet at the rate of 3 to 4 percent of its total grain ration to encourage him to drink more water.
- Adding ammonium chloride to the diet to acidify the animal's urine. This renders crystal components more soluble, thus more likely to be expelled in the urine before they form stones. Preventive doses of ammonium chloride can be added to grain rations at the rate of one level teaspoonful per 150 pounds of goat, divided into two daily feedings. Be forewarned: Ammonium chloride isn't very tasty, so you may have to conceal it in a carrier such as a dab of molasses or flavored yogurt. Hoeggers Goat Supply and Molly's Herbals (see Resources) sell it by the pound.
- Because castration of young kids removes the hormonal influence necessary for full development of the urinary tract, it's wise to allow future working wethers' urethras to mature at least partially before castrating the animals, at four to six months of age.

If you suspect a buck (or wether) has a stone, contact your vet without delay; treatment must begin as soon as possible if the goat is to be saved. Treatments of choice are:

- **Surgical removal of the urethral process.** This doesn't affect an animal's future as a stud buck and it restores urine flow in about a third of early cases. In extreme emergencies, when no vet is available, producers have been known to use sharp scissors to snip off an afflicted goat's pizzle themselves; it's not the ideal protocol, but it just might save a buck's life.
- **Administration of acepromazine maleate (Ace).** This common animal tranquilizer, or flunixin meglumine (Banamine), an anti-inflammatory painkiller, may counteract spasms of the urethra and relax the muscles surrounding the penis enough to allow minor stones to be passed.
- **Administration of 2 tablespoons of ammonium chloride dissolved in 60 cc of water.** For minor blockages, a two-week course is sometimes effective.
- **Anesthetizing the goat and tapping its bladder.** This may remove accumulated urine. (Note that this isn't a cure; it's a procedure to buy time.)
- **Surgery to manually drain the bladder** and remove urinary stones.

Keep in mind that a goat that successfully passes one stone probably has more where that one came from and is at considerable risk to make more. The stone should be taken to the vet, who will submit it to a testing laboratory to determine which of several possible mineral combinations have formed it. Based on those findings, the vet will formulate a treatment and prevention plan for all of the male goats raised on your farm.

Those Does

There are few sights prettier than a pasture full of your own beautiful does and their happy, healthy kids. Putting those kids in the picture is every goat producer's goal; here are some strategies to achieve this goal.

- **The brood does you pick for your breeding program must be healthy and productive easy keepers.** A herd of big fat mama goats that eat like elephants and produce single kids once a year will drain your wallet faster than you can say "cull doe." Review chapter 5 *before* you go goat shopping!
- **Breed them to the right kind of buck.** Choose a buck to complement your does and your breeding program. Buy one from a litter of two or three, not from a dam that consistently produces single kids. Check his teat structure; he has as much influence on his daughters' teat structure as do their dams. Don't expect your does to carry the load; remember the old adage "The buck is half the herd."

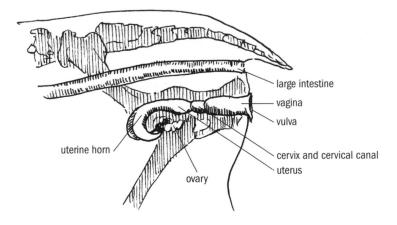

A DOE'S REPRODUCTIVE SYSTEM

- **Feed them correctly.** Dry (open; nonpregnant, nonlactating) does don't need grain. If their pasture is depleted, feed good grass hay and protein blocks. Make certain they have a plentiful supply of clean water and a properly balanced mineral available at all times. Three weeks before breeding, "flush" them by adding an increasing amount of grain to their diet until each doe is consuming one to two pounds of grain per day. Flushing encourages multiple ovulation, which often results in larger litters of kids per doe. After breeding, cut out the grain until three or four weeks before kidding, then introduce it again, working each doe up to a pound to a pound and a half of good grain mix per day.
- **Confirm pregnancy and see how many kids your does are carrying by using ultrasound.** It's effective and inexpensive, and it allows you to divide your does into groups and feed them according to their breeding status; a doe pregnant with quads requires more feed and closer management than one carrying a single kid.
- **Vaccinate and boost with Bo-Se.** In addition to their annual CD/T or Covexin-8 and other herd-specific boosters (we use Covexin-8, Case-Bac for CL, and Super Poly Bac for pneumonia on our farm), boost your does again at three to five weeks before kidding so they're sure to pass antibodies against disease to their kids via their colostrum. A kid's own immune system doesn't begin functioning until it's 8 to 10 weeks old, so it needs that passive transfer of immune bodies from its dam to stay healthy. At the same time we boost our does, we inject them with Bo-Se, a selenium and vitamin E supplement. This is a common practice in selenium-deficient parts of the country; check with your vet or county Extension agent to see if it's needed in your locale.
- **Be there when they kid.** Say this aloud three times; make certain it's imprinted in your mind. Too much can go wrong to leave kidding to chance. We'll talk about this again later in this chapter.
- **Protect those kids from predation.** Even a fox can carry off a newborn kid. Use your best guardian animals in pastures where neonatal kids are present, or keep these kids penned near the house until they're bigger.
- **Monitor kids' health at least twice a day.** It doesn't take long for a kid to die from coccidiosis, enterotoxemia, or a half-dozen other life-threatening kidhood diseases.

You've Got to Be Kidding

The highlight of every goat rancher's year is kidding time. It's exhilarating, terrifying, and profoundly satisfying, all at the same time.

Preparing for Kidding

Prior to your first doe's kidding, assemble a complete kidding kit or check the one you already have, making certain everything you need is present and accounted for. Set up one or more kidding areas (in the barn or in easily accessible paddocks or pens) and jugs (mothering pens where each doe can bond with her kids for two or three days before rejoining the herd), depending on the number of does you'll kid out. If you don't use homemade milk formula or have a lactating dairy goat to provide milk for bottle babies, buy a bag of your favorite milk replacer to have on hand.

A Well-Stocked Kidding Kit

We pack our kidding supplies in a Rubbermaid toolbox-stool. It's roomy, it has a lift-out tray for small items, it's easy to tote to the barn, and it's so much nicer to sit on than an overturned five-gallon bucket. The box holds:

- A bottle of 7 percent iodine (used to dip newborns' navels; you'll have to get it from a vet because the United States Drug Enforcement Agency regulates sales of all products containing more than 2.2 percent iodine.)
- A shot glass (to hold the navel-dipping iodine)
- Sharp scissors (for trimming umbilical cords prior to dipping them in iodine), always stored disinfected and in a sturdy zip-top plastic bag
- Dental floss (to tie off a bleeding umbilical cord if necessary)
- A digital veterinary thermometer
- A bulb syringe (the kind used to suck mucus out of a human infant's nostrils)
- A rubber leg snare (cord snares work well too)
- Shoulder-length obstetric gloves
- Two large squeeze bottles of lubricant (we like SuperLube from Premier1)
- Betadine scrub (for cleaning does prior to assisting)
- A sharp pocketknife (you never know when you'll need one)
- A hemostat (ditto)
- A lamb- and kid-carrying sling
- An adjustable rope halter

- Two flashlights (we like to have a backup in case the first flashlight fails)
- Two clean terry-cloth towels

Is This Doe Ready to Kid?

A doe's gestation period lasts between 145 to 155 days; 150 days is said to be the norm. Does don't always go by the book, however, so you'll need to watch for signs of impending kidding as each doe's big day approaches.

Udder Fill

The average first-time doe's udder will begin filling about a month before kidding; veteran does fill anywhere from a few weeks to immediately prior to delivering their kids.

ARE YOU READY FOR KIDDING?

- Though most does kid without assistance, be ready to help out if the need arises. Keep your fingernails clipped short and filed throughout kidding season (you won't have time for a manicure when an emergency occurs).
- Know how to determine what configuration a malpositioned kid is in and precisely how to correct it. If you think you might forget, photocopy and laminate instructions to keep in your kidding kit.
- Practice. Teach your fingers to "see." Borrow a pile of plush toy animals and put one in a fairly close-fitting paper bag. Without looking, stick your hand in the bag and figure out what you're feeling. Switch animals. Add another animal to the mix, then yet another; see if you can sort out triplets.
- Program appropriate numbers into your cell phone so you can call the vet, your goat mentor, or your choice of Goat 911 responders at the touch of a button.
- Install a baby monitor in your barn. Does thrash around and cry when they go into labor; an inexpensive audio monitor saves many moonlight strolls to the barn (video models are even nicer).

Right before kidding, most does develop strutted udders (see illustration on page 84). A strutted udder is so engorged with milk that it's shiny and its teats stick out to the sides somewhat. If the goat has light-colored skin, her strutted udder will take on a rosy glow.

Softening Pelvic Ligaments and Other Physical Changes

One of the best ways to predict impending kidding is by monitoring the pelvic ligaments in a doe's hindquarters. The ligaments in question attach at either side of the spine about midway between the hips and pin bones, then they angle away toward the rear and away from the spine (one authority suggests they take on the configuration of a peace sign). To locate these ligaments, slide your hand along the doe's spine, including the area an inch to either side. They're about as big around as a pencil and very firm — until secretion of the hormone relaxin softens them in preparation for giving birth. These ligaments grow increasingly softer as kidding day approaches. When you can't feel them any longer, expect kids within 24 hours.

Relaxin causes other structures in the pelvis to relax as well. This results in the rump becoming increasingly steeper as labor approaches, both from hips to tail and as seen from side to side. As this happens, the area along the spine seems to sink and the tailhead rises. About 12 hours before kidding, you can grasp the spine at the tailhead and almost touch your thumb to your fingers on the other side. It's a dramatic change and one you won't believe until you've felt it firsthand.

The perineum, the hairless area around the vulva, sometimes bulges during the last month of pregnancy. About 24 hours before kidding occurs, the bulge diminishes and the vulva becomes longer, flatter, and increasingly flaccid.

As the cervix begins to dilate, the cervical seal (wax plug) liquefies. When this occurs, does often discharge strings of mucus from their vulvas ranging from a clear, thin substance, to one that's thicker and opaque white, and, finally, to thick, amber-colored discharge tinged with amniotic fluid.

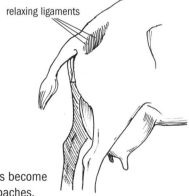

relaxing ligaments

The ligaments in a doe's hindquarters become increasingly relaxed as kidding approaches.

Did You Know?

- Single-born bucklings are generally born a day or two sooner than single-born doelings.
- Multiples are usually delivered two or three days earlier than singles.
- When twins of opposite sexes are born, the male is almost always delivered first.
- Many does prefer male kids. If one kid from a litter is rejected, it's usually a doeling.

Was That a Labor Pain?

During first-stage labor, relatively mild uterine contractions cause a doe to pause for a moment to stretch and raise her tail or to lie down briefly and hold her breath for a heartbeat or so.

As second-stage labor begins, most does lie down on one side, extend their hind legs, and cry out in pain as long, hard abdominal contractions wrack their bodies.

Behavioral Changes

The following behaviors indicate that the doe is in first-stage labor, which generally lasts 12 to 36 hours. During this time she may:

- Drift away from the herd to seek a nesting spot, sometimes in the company of her dam, a daughter or sister, or a best friend.
- Engage in nesting behavior by digging a depression in her stall or pen, lying down, getting up, circling, digging, and repeating the cycle over and over.
- Gaze off with a wide-eyed, unblinking, faraway look on her face.
- Diligently search for something (her unborn kids) while "talking" in a soft, low-pitched voice.
- Become unusually affectionate — or standoffish.
- Yawn and stretch (stretching helps put her kids into birthing position).
- Pant, taking short, shallow breaths.
- Walk very loosely on her hind legs (as hormonal changes relax her pelvis, her rear-leg angulation alters, affecting her gait) and move more slowly than usual.
- Stop eating and/or grind her teeth (indications that she's in pain).

Get Right Down to the Real Nitty-Gritty

Eventually, second-stage labor begins. The doe lies down and rolls onto her side when a contraction hits, rides out the contraction, then rises and repeatedly repositions herself until she finds a spot she likes. Once she does, she may roll up onto her sternum between contractions, but she'll usually remain lying down. Some does deliver standing up or in a squatting position, however, and that's normal too.

The first thing to appear at the doe's vulva is a fluid-filled, water balloon–like sac called the *chorion*, one of two separate sacs that enclose a developing fetus within its mother's womb (the other is the *amnion*). Either or both sacs can burst within the doe or externally as the kid is delivered.

In a normal front-feet-first, diving-position delivery (and this occurs about 60 percent of the time), a hoof appears inside the chorion (or directly in the vulva if the chorion has already burst), followed by another hoof, and then the kid's nose, tucked close to his knees. Once the head is delivered, the rest of the kid quickly follows.

In a normal hind-feet-first delivery, two feet appear followed by hocks. Because the umbilical cord is pressed against the rim of the pelvis during this delivery, it's wise to *gently* help the kid out once his hips appear. Otherwise, both are generally textbook deliveries and rarely require assistance.

The following presentations contribute to *dystocia* (difficult or abnormal delivery). Sooner or later (and it's bound to be sooner than you want it to be), you'll have to help a malpositioned kid into the world. When it happens, keep these points in mind.

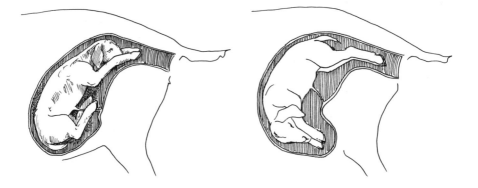

Normal diving-position delivery   Normal hind-feet-first delivery

- Because of anatomical differences, you can't pull a kid with tackle and a great deal of force, the way you'd pull a calf. If you must pull, use lots of lubricant and pull while the doe is having a contraction. If her contractions have stopped, exert no more pressure than a man can apply with one arm. Don't pull straight back; pull out and down in a gentle curve toward the doe's hocks.

- If you've purchased a kid snare (made of rubber), be aware that you don't use this appliance actually to *pull* a kid. It's only supposed to fasten around legs so that you don't "lose" them while correcting dystocia. If you pull, the snare stretches; this is not a good thing. To pull a kid, simply grasp the legs, preferably above the pasterns but below the knees, then pull. In an emergency, you can pull the head too.

- Before entering a doe, make certain your fingernails are short and that you've removed your watch and rings. Wash her vulva using warm water and mild soap or an antimicrobial product such as Betadine scrub. Pull on an obstetric glove if you have one; if you don't, scrub your hand and forearm with whatever you use to clean the doe and then liberally slather the glove or your hand and arm with lubricant. Now, pinch your fingers together and gently work your hand into the vulva. The goat will probably scream (and possibly keep screaming the entire time you're helping her), so don't be surprised when she does.

- Determine which parts of the kid are present in the birth canal. Closing your eyes and moving your awareness to your fingertips can help. Follow each leg to the shoulder or groin if you can, making sure the parts you're feeling belong to the same kid. When you're certain they do, if you can manipulate the kid into a normal birthing position, do so, and then help pull out the kid (by this time, the doe will definitely be too exhausted to do it by herself). If you can't reposition the kid in ten minutes, stop trying and call your vet for assistance.

- The inside of a doe is extremely fragile and if you or the kid tears her, she's likely to die. When repositioning the kid, cup your hand over sharp extremities such as hooves and work carefully and deliberately; her life depends on your gentle technique.

- If you have to pull one kid, pull them all. The doe will be exhausted; simply help her get them out so she can rest.

- Any time you go into a doe, you *must* follow the birth with a course of antibiotics. Ask your vet for specific directions.

The first kid should be born within an hour after hard labor commences. Many producers swear by the 30×3 rule: Allow 30 minutes after hard labor begins for the birth membranes to appear, another 30 for the kid to be born, and 30 more minutes for second and subsequent kids to be delivered. If things aren't resolved in that time frame, you'll have to assist or call the vet.

Dystocia and How to Help

There are a number of ways you can intervene to help a doe who's dealing with dystocia, and assisting isn't as hard or as scary as it seems to be. Following are the situations you're most likely to encounter.

True breech presentation (rear end first, legs tucked forward): Some does can give birth in this position and others can't. If yours isn't able to, call your vet. It's best not to try to help correct this positioning yourself. If you must reposition a breech kid, however, try to elevate the doe's hindquarters before you begin. Push the kid forward, work your hand past its body (it'll be a tight squeeze), and grasp one hock. Raise up the hock and rotate it out away from the body. While holding the leg in this position, use your little and ring fingers of the same hand to work the foot back and into normal position. Repeat on the opposite side. The umbilical cord will be pinched, so pull the kid as quickly as you safely can.

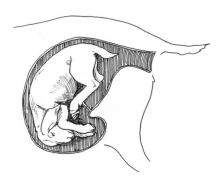

True breech presentation

One leg back: Many does can deliver kids in this position, but if yours isn't able to, push back the kid just far enough to allow you to cup your hand around the offending hoof and gently pull it forward.

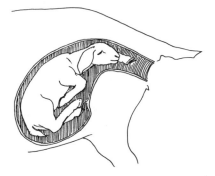

One-leg-back delivery

Head back: A small kid with its head bent back to its side can sometimes be pulled, but do your best to correct its position before doing so. To correct, attach cords or a rubber kid snare to the front legs so you don't lose them, then push back the kid as far as you can and bring the head around into position. Alternate difficult presentation: Sometimes the front legs are presenting but the head is bent down. This is more difficult to correct. Attempt to correct it in the same way as you would if the head was bent back, but call the vet to be sure he or she is driving to your place before you commence.

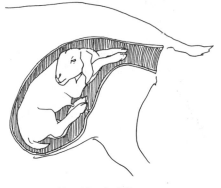

Head-back delivery

Crosswise: This is a bad one, so call the vet immediately. In the meantime, push back the kid as far as you can (elevating the doe's hindquarters will help) and determine which end is closer to the vulva, then begin manipulating that end into position. This kid is usually easier to deliver hind feet first.

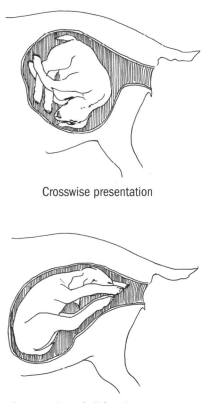

Crosswise presentation

All four legs at once: Attach cords or a rubber kid snare to a set of legs, making certain you have two of the same kind, and then push back the kid as far as you can. Reposition the kid for either a diving-position or hind-feet-first delivery, depending on which set of legs you've captured.

Presentation of all four legs at once

Twins coming out together: Attach cords or a kid snare to two front legs of the same kid (follow the legs back to their source to make sure they're attached to the same kid). Push back the other kid as far as possible and bring the captured kid into the normal birthing position.

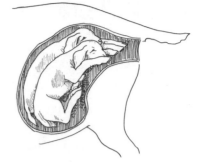

Twins coming out together

Twins coming out together with one reversed: Follow the same protocol as for twins coming out together (above), but because it's generally easier to do so, pull the reversed kid first. If both are reversed, pull the kid closer to the vulva.

To help you prepare to assist your does in giving birth, I strongly suggest you visit Fias Co Farm's kidding image galleries (see Resources) and study the material at length. If "a picture is worth a thousand words," this Internet resource is a virtual encyclopedia!

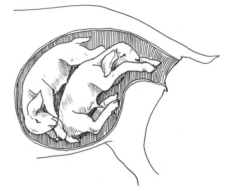

Twins coming out together, one reversed

There Be Kids!

When the first kid arrives, remove birthing fluids from his nose by stripping your fingers along the sides of his face. If he's struggling to breathe, use the bulb syringe from your kidding kit to suck out the fluid. If he's really struggling to breathe or isn't breathing at all, take a secure grip on his hind legs between the hocks and pasterns, place your hand behind his neck (near the withers) to support it, and swing him in a wide arc to jump-start his breathing.

It's worth mentioning for beginners' sakes that an alarming number of assisted kids seem to be born "dead." Actually, their hearts are pumping but they haven't started to breathe. Swinging usually gets them going; tickling the inside of their nostrils with a piece of straw or hay works well too. Don't give

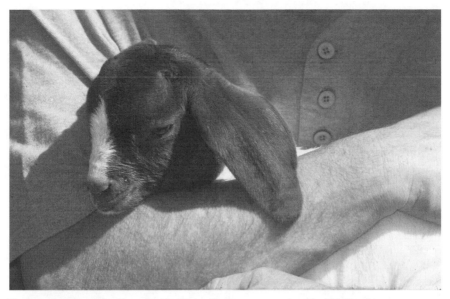

The successful culmination of your doe's efforts: one more healthy kid, like this Boer doeling.

up. If you keep stimulating these kids, chances are they'll start to breathe and be perfectly all right.

Once he's breathing, place the kid in front of his dam so she can begin cleaning him. It's the taste and scent of her kids that creates her maternal bond. At some point she may leave him to deliver another kid. This is normal. Simply place both kids in front of her after the second has arrived.

Once all are accounted for (if in doubt, go in and check), think *snip, dip, strip,* and *sip. Snip* the umbilical cord to a manageable length if it's broken off overly long (about an inch and a half is just right). *Dip* the kid's cord in 7 percent iodine (fill a shot glass, film canister, or dairy cow teat dip cup with iodine, hold the container to the kid's belly so the cord is completely submerged, then tip the kid back to effect full coverage; don't omit this important step). *Strip* the doe's teats to make certain they aren't plugged and that she indeed has milk, then make sure the kids *sip* their first meal of vitally important colostrum within an hour or so after they're born.

When Things Go Amiss

Unfortunately, the sweet birth scenario doesn't always play out to the end. The doe dies, she rejects one or all of her kids, or a kid is too weak to nurse from its dam.

THOSE IMPORTANT FIRST MEALS
OF COLOSTRUM

Colostrum is a thick, yellowish milk does produce for about 48 hours after kidding. It's jam-packed with important nutrients, but more important, colostrum contains antibodies that kids need in order to survive.

The lining of a kid's intestine can absorb those antibodies for roughly 12 to 24 hours after birth. Kids that don't ingest colostrum lack immunity to serious diseases such as enterotoxemia, tetanus, and *E. coli* until their own immune systems begin to function. Without it, kids rarely survive: colostrum is that important.

- If a kid can't nurse from his dam but she's still available, milk her and tube- or bottle-feed her colostrum to the kid in two- or three-ounce portions until it has ingested 10 percent of its total body weight in fluid.
- If you can't use colostrum from a kid's dam, fresh or frozen colostrum from another doe will do just as well. Sheep colostrum, especially from sheep kept on the same farm, works well too. Cow colostrum runs a distant third place.
- You can freeze colostrum for up to one year. Quick-freeze it in two- or three-ounce feedings in small-portion human infant bottles or

Tube Feeding

Healthy kids are born with a strong suck reflex, and if the mother isn't available, they can immediately be started on a bottle. Premature or other weak kids may have to be fed through a stomach tube, at least until they're stronger and can suck.

Most everyone approaches a first tube-feeding session with fear and trepidation. Don't worry; the process is easier than you think.

You'll need the proper equipment: a feeding tube and a 60 cc syringe with the plunger removed. Soft plastic tubes designed specifically for the job work best (find a list of suppliers in Resources), but an open-ended, 6-millimeter-diameter piece of 20-inch-long plastic tubing (ask for it at the hardware store) will also do in a pinch. Always sanitize the feeding tube and syringe between in double-layered, zip-closing plastic sandwich bags. Avoid storing it in self defrosting freezers; constant thaw and refreeze cycles affect its integrity. Never microwave colostrum; this kills the protective antibodies. Instead, immerse the container of the liquid in a hot-water bath until the colostrum registers 102 to 104 degrees (measure it to be sure).

■ If no colostrum is available, a kid may survive if promptly given a weight-appropriate injection of an immunoglobulin (IgG) replacer such as Goat Serum Concentrate (you can buy it from Hoegger Goat Supply; see Resources) or a bovine IgG replacer such as Seramune Bovine IgG (which large-animal vets are more likely to have on hand). These are *not* the same thing as the inexpensive, powdered supplements based on cow colostrum that you'll find on the shelf at your local feed store. If you have nothing else, try them, but most who do report very limited success.

■ Colostrum-deprived kids (as well as kids from unvaccinated dams) should be given CD and tetanus antitoxin injections (two shots; don't combine them) within a day after birth and at two-week intervals until they're 12 weeks old. They should also be exposed to disease as little as possible; keep them housed away from other livestock.

uses and warm the tube just prior to use by immersing it in a bowl of clean warm water.

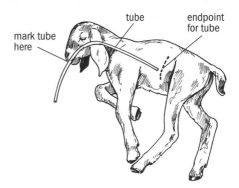

1. To tube-feed a kid, first place the tube alongside his body with one end at the animal's last rib, as shown in the diagram. Mark the tube near the animal's mouth.

2. Sit in a comfortable position with the kid facing away from you, its shoulders restrained between your knees and its body dangling down between your legs. The goal is to insert the tube into the kid's esophagus, not the trachea (windpipe), and with its body in this position, the tube is almost certain to go where it's supposed to.

2a. Place the kid between your knees, looking out.

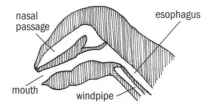

2b. Windpipe and esophagus when a goat swallows

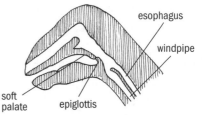

2c. Windpipe and esophagus when a goat breathes

3. Take the tube out of the warm water, insert the rib end in the kid's mouth, and slowly feed the tube down the baby's throat. In this position, he will readily swallow the tube.

Feed the rib end of tube into the kid's mouth.

4. Check to make certain the tube is in the esophagus, not the trachea. Here's how you can tell it's in the esophagus: (1) As you insert the tube, you can see it advance along the left side of the kid's neck; (2) the kid shouldn't gag or cough (although it may briefly struggle); (3) the tube can be easily inserted all the way to the mark you made on it (a tube inserted into the trachea can't advance that far).

5. Attach the mouth of the feeding tube to the empty syringe. Fill the syringe with warm fluid and allow it to gravity-feed into the kid's stomach.

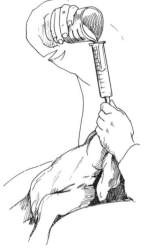
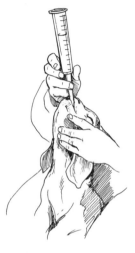

5a. Attach the tube to the syringe.

5b. Fill the syringe.

5c. Gravity-feed the kid.

6. Next, detach the syringe and crimp off the end of the tube, then quickly and smoothly withdraw the tube (don't allow fluid to leak from the tube as you pull it out).

Detach the syringe and crimp the end of the tube.

You may have to tube-feed a sick or weak kid for several days until it begins to suck. You can also tube-feed warm electrolyte solution to scouring kids.

AN INEXPENSIVE START IN QUALITY GOATS

Many goat ranchers prefer not to raise orphans and weak kids that need to be bottle-fed. They often give them away or sell them at rock-bottom prices. If you have more time and energy than money and you'd like to improve your herd, put out the word that you're interested in a well-bred bottle-baby buck. Call or e-mail producers within driving distance from you and post your interest at breed-specific e-mail lists. We've obtained and raised two high-quality bucks this way.

Bottle Feeding

Goat ranchers either love or hate bottle babies. I reside firmly in the former camp. It's expensive and time consuming to raise kids on a bottle, but for some of us, it's worth it!

What to feed: The best food for baby goats is goat milk. If you own many meat goat does, consider buying a dairy doe to produce milk for the bottle babies you're bound to produce (for more information, see appendix E, A Milk Goat for the Kids). Barring that, consider mixing a formula based on homogenized full-fat cow's milk from the grocery store. Two good formulas are: (1) four parts milk to one part half-and-half; and (2) one gallon milk mixed with one 12-ounce can of evaporated milk (*not* condensed milk) and one cup buttermilk. Remove enough milk from the gallon of whole milk to add the evaporated milk and the buttermilk. Mix well, then add back as much of the remaining whole milk as you can to make a full gallon.

If you must use powdered milk: As a replacement for fluid milk, choose a product based on milk protein, *not* soy (baby ruminants have trouble digesting soy), and be sure it's one designed specifically for feeding baby goats. If kid milk replacer isn't available, seriously consider one of the homemade milk-based formulas or substitute a quality lamb milk replacer diluted with extra water (and if the kid develops diarrhea, add still more water).

How to feed: Some people may tell you it's okay to feed neonatal kids only two or three times a day. Don't believe them! For the first few weeks, you *must* feed kids frequently. Here is the schedule we use, based on feeding Boer goat kids.

- Days 1 and 2: Two to 3 ounces of colostrum per meal, with a meal fed every 4 hours
- Days 3 and 4: Three to 5 ounces of goat milk or homemade formula per meal, with a meal fed every 4 hours
- Days 5 through 14: Four to 6 ounces per meal, fed every 4 hours
- Days 15 through 21: Six to 10 ounces per meal, fed every 4 hours
- Days 22 through 35: Gradually work up to feeding 16 to 20 ounces per meal 3 times per day
- At 10 weeks of age, we eliminate the noon meal (by then the kid is eating plenty of grass or hay and nibbling at a small ration of commercial goat feed).
- At 12 weeks, we cut back the meals to 8 ounces each.
- At 14 weeks, we wean the kid.

WHEN BABY WON'T EAT

There are few things as frustrating as kids that won't accept a bottle. These are generally kids that nursed their dams and are pretty sure the nipple you offer them is poison. Here are some strategies to consider in such cases:

- Switch nipples until you find one the kid likes. They do have preferences — unfortunately.
- Lightly coat the nipple with fruit yogurt or sugar.
- Tickle the baby's rectum as you offer him his bottle (does nose their babies' bottoms while they nurse).
- Cup your hand above the baby's eyes to simulate the darkness of Mama's groin.
- Offer him a bottle and if he won't take it, put him away. It may take three or four of these sessions until he's finally hungry enough to eat.
- Persevere. In the meantime, tube-feed the kid if necessary. He *must* stay hydrated in order to survive.

To start a kid on a bottle, sit on the floor with your legs crossed and the kid sitting in your lap on its rear end, facing away from you. Open its mouth with your left hand, insert the nipple, and, with your palm under the kid's jaw, use your fingers to keep the nipple aligned with its mouth. With your right hand, elevate the bottle just enough to keep milk in the nipple, adding more tilt as the kid empties the bottle.

If the kid doesn't suck, gently squeeze the bottle so a *tiny* amount of milk goes into its mouth. Don't drown it!

Some newborns take to the bottle right away. Others struggle and fuss. Keep trying. The little one has got to eat or it'll die.

Feeding gear: Every producer has his or her favorite bottle-feeding gear; we prefer Pritchard teats attached to plastic soda bottles. The genuine Pritchard teat (there are imitations — beware; they aren't as good) is an oddly shaped red nipple attached to a yellow plastic cap ring sized to fit standard screw-top household containers (including soda bottles). Its plastic base incorporates a "valve" that makes fluids flow smoothly through the nipple.

We like soda bottles because they're readily available and after a day's feedings we can add them to the recycling bin and start afresh.

After each feeding, wash the nipple and bottle in hot, soapy water and thoroughly rinse them. If you use standard bottles, sanitize them once a day in a solution of one part household bleach to ten parts water, and then very thoroughly rinse them. But don't bleach nipples of any kind; bleach tends to degrade them very quickly.

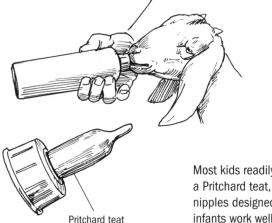

Pritchard teat

Most kids readily nurse a Pritchard teat, though nipples designed for human infants work well too.

A KID IN THE HOUSE

If you have only one or two bottle babies to feed, save yourself those midnight trips to the barn by raising them in the house. Ours spend their time in the living room in an extra-large, top-opening dog crate. Bedded on frequently changed pieces of old blankets, they're practically odor-free, extremely handy to care for, and more entertaining than a TV sitcom.

If you really have a sense of humor, buy cheap disposable diapers for human infants, trim a little hole for their tail, and release your kids for playtime in the house.

Tips and observations:

- Block off or screen all power cords. Computer and telephone cords appear to be especially attractive to kids.
- Move pet food out of reach and don't allow kids to use your dog for a trampoline.
- Bathtubs are very attractive; kids use slick surfaces to exercise by running in place
- Once kids begin to climb, *watch them* while they're loose in the house. They'll leap from one object to another and end up in the darndest places (on the top of the refrigerator, for instance).
- Goats eat paper. That means toilet tissue, money, and your two-legged kids' homework. Keep all paper out of reach.
- There's no better ending to a tiring day than snuggling with a cuddly baby goat and reading the latest copy of *The Boer Goat*.

Producers Profile

Beth and Randy Ellerbrock and Spencer Toner, Star E Ranch
Clayton, Illinois

Beth and Randy Ellerbrock and their son, Spencer Toner, raise show-quality Boer goats and quality Quarter Horses. Randy, a truck driver and team roper, and Beth, a part-time cardiac rehab nurse, purchased their first Boer does in 2004. By researching the breed before buying and starting with quality stock, then learning to prepare and handle it correctly, they're already making their mark in the show ring, proving that newcomers can shine in the show goat industry.

Q *Beth, what sparked your interest in meat goats?*

A Randy and I married in 2003. I was a city girl who loved animals and wanted horses, but I'd never really seen or had anything to do with goats except the ones in petting zoos. One day Randy took me to the sale barn — I'd already talked him into a bucket calf — and we came home with a Pygmy cross with two kids at her side. They were wild and wouldn't let me touch them. I was highly disappointed! So a few weeks later we bought Archie, a Nubian bottle baby. He lived in the house for two and a half months. He was awesome, but he was killed when he got in with the horses. By then we'd decided that goats were pretty neat, but we wanted a *breed*, so we did some research and thought Boers were interesting — they were big and docile, and a good profit could be made raising breeding stock and show stock, so we went into Boer goats full steam ahead. We bought our first fullblood doe in June of 2004 and as of today we have about 90 goats on the ranch. Most are fullbloods, a few are percentages, and some are dairy does we use as wet nurses to raise orphan kids.

Q *Are you happy with Boers?*

A Yes, especially since getting involved in showing. There are so many shows to go to with this breed.

Q *Is there a good market for goats in Illinois?*

A We currently have the only registered herd in our county, and we focus on breeding and show stock, but with every herd, you have to have a place to sell your culls. In just one year our local sale barn has gone from having only a handful of goats sold per week to 30 or so at each sale, and a good percentage of these are now Boer crosses. Several meat buyers frequent our sale barn and that keeps the prices up. I've noticed people are bringing goats from farther away now because of this.

Q *Do you think there's room for new producers in this industry? Is it generally harder or easier for them to start in meat goats than it was for you?*

A The United States is still importing more goat meat than we can produce, so there's a lot of room for new producers — and there's always room in the show ring for friendly competition!

I think anyone new to meat goats in our area would have it a little easier than we did. First off, we now have the local vets learning more about goats. Prices are steadily rising at the sale barn due to demand, so that's also a help. We're also seeing real progress at our local county fair. They no longer think the goat classes are for just dairy goats, and this year's number of entries almost doubled from two years ago.

Q *What advice would you give someone interested in raising show-quality meat goats?*

A Go to a few goat shows and watch. Listen to the judges, talk to the breeders, pick up business cards. Listen and learn. Then start small — don't buy a top-of-the-line show doe right away; buy a nice percentage herd or a few fullblood does and a good fullblood sire. You can add great genetics to your herd without spending a lot of money if you use AI [artificial insemination]. The best bucks in the country are selling for over $25,000, but for $150,000 you can have your doe artificially inseminated by one of these top bucks. That's a huge savings! Also, there's a huge learning curve in health care issues. You'll lose a goat or two along the way, so it's better to start with somewhat less expensive goats.

14

Marketing
Meat Goats

THIS CHAPTER TAKES A MORE IN-DEPTH LOOK at ways you can market meat goats. Certainly, you can market goats (as chevon, or goat meat) raised for slaughter either organically or otherwise, but in addition to this avenue for meat goat producers, goats can be raised for breeding, for showing, and even for working. As goat popularity rises, markets both for their meat and for live animals themselves will continue to expand.

Market Goats

Most goat ranchers produce market goats (also called slaughter goats) ranging in age and size from milk-fed kids to meaty young goats bred for ethnic-holiday consumption. Older culls enter the meat market too, often selling to Caribbean islanders, who prefer the robust flavor of meat from mature goats. (See the chart on page 14.)

Whether you sell at auction, through intermediaries, or straight from your own back door, to succeed in the meat goat business you must constantly strive to meet your customers' needs. The best way to learn about the opportunities in your area is to join a state or regional meat producers group and pick the minds of established goat ranchers.

Goat producers sell market goats in a variety of ways, depending on available options and personal preference. These include but aren't limited to:

A MARKET GOAT VOCABULARY

Shoebox kids. Newborn to ten-day-old kids. These are generally excess dairy-breed males that are bottle-fed colostrum for a few days and then taken to auction, where they're frequently purchased to be used as pets, but they sometimes sell to feeder operations too.

Suckling kids (Easter kids, hothouse kids, cabrito, capretto, katsikia). Plump, unweaned, milk-fed kids four to six weeks old. The preferred size for cabrito (sold to Hispanic buyers) and capretto (Italian market) kids runs 18 to 25 pounds live weight; katsikia (Greek market) kids are heavier at 30 to 40 pounds.

Market kids. Weaned kids with milk (baby) teeth. These are further differentiated as *feeder kids* (lean kids in demand by feedlot producers), *Muslim kids* (moderately lean kids preferred by ethnic buyers, who generally will not accept fat goats), *finished kids* (plump but not fat older kids), *yearling kids* (young goats between one and two years of age, as judged by their teeth), and *culls* (young cull does are sometimes called Philippino goats; older does and bucks may be referred to as curry goats).

- **Auction sales.** They're the easy way to market your goats. You're nearly always assured a buyer and you're paid by a bonded entity on sale day. On the other hand, you have little or no control over prices, and if you don't accept the one you're offered, you must pay a no-sale fee and take home goats exposed to sale barn diseases. To find an area sale barn that handles meat goats, ask local goat producers for their recommendations or pick up the phone and call around. Special sheep and goat auctions are better than anything-goes livestock sales because more high-volume meat buyers attend them. Whether selling or simply observing at such auctions, join in conversations and hand out your business card. Networking often returns big dividends.
- **Livestock dealers and brokers.** Dealers come to your farm to buy your goats. Livestock brokers are dealers who handle your animals on commission.
- **Meat packers.** Packers operate slaughterhouses. Some pick up at the farm, while others expect producers to deliver goats direct to their facilities.

- **Direct meat sales.** All meat sold to restaurants, meat markets, and individuals must be processed at federally inspected USDA slaughterhouses. There are stringent laws regulating the sale of processed meat, so investigate legalities before you proceed.
- **Sales of live meat animals to consumers.** These fall under one of two classifications: With *on-farm sales,* customers come to your farm, pick out a goat, and either take it with them or slaughter it on-site. Many states restrict on-site slaughter, so find out the particulars before you allow it. With *freezer sales,* customers select a goat or arrange for the purchase long distance. You then deliver the animal to an approved slaughterhouse, after which the customer takes delivery of the packaged meat.

For an introduction to these and other options, read "Marketing Slaughter Goats and Goat Meat," by Titania Stanton, Module 7, in Langston University's web-based Training and Certification Program for Meat Goat Producers (see Resources).

Organic or Natural Chevon

While most goat meat is produced for ethnic markets, a second, smaller segment of the American public is turning to goat meat for its health-giving qualities. Because naturally lean chevon doesn't marble fat the way other red meats do, external fat can be trimmed, producing low-fat, low-calorie, high-protein cuts of meat.

Because the general American palate isn't attuned to strong-flavored meat, producers planning to market goat meat to mainline consumers must

CALORIE, FAT, AND PROTEIN CONTENT OF VARIOUS KINDS OF MEAT

Roasted Meat (3 oz)	Calories	Fat (g)	Saturated Fat (g)	Protein (mg)
Goat	122	2.58	0.79	23
Beef	245	16.00	6.80	23
Pork	310	24.00	8.70	21
Chicken	122	3.50	1.10	21
Lamb	235	16.00	7.30	22

Source: USDA Handbook No. 8, 1989.

PROMOTE YOUR PRODUCT

If you want to sell more goat meat, it's important to promote chevon ("the healthy red meat") wherever and as often as you can. Because most Americans have never tasted goat meat, they think they won't like it, so it's up to you to show them that they're wrong. Try distributing roasted or barbecued chevon on toothpicks at fairs, health food stores, and food co-ops. Hand out easy recipes and cooking instructions to those who express interest; it's important that their first culinary adventures with goat meat don't flop. Carry a chevon dish to your church's potluck dinner, and serve it at home to your friends. Don't tell diners they're sampling goat meat until they've eaten it. They're apt to be pleasantly surprised.

produce a mild-tasting product and teach buyers how to prepare it correctly.

Although opinions vary, most mainstream goat meat connoisseurs prefer meat from meat goat breeds to that of dairy goats. Furthermore, meat from young goats is generally chosen over dishes prepared from the meat of mature animals.

Organic — or Not?

During the 1990s, organic production took U.S. agriculture by storm. According to the USDA's "Organic Farming Statistics and Social Research," published by the USDA's Economic Research Service and based on the 2002 U.S. Census of Agriculture, the number of acres of certified organic cropland quadrupled between 1992 and 2001. The number of certified beef cows, dairy cows, and pigs also quadrupled during the same fiscal period.

Because of America's burgeoning interest in all things organic, prospective goat ranchers frequently decide to produce organic chevon. Few, however, go on to raise organic goats. Here's why: In 1990, Congress passed the Organic Foods Protection Act and, in doing so, created the National Organic Program (NOP) and the National Organic Standards Board (NOSB), from which emerged a single set of standards for organic production, labeling, and marketing, called the National Organic Standard. All foods grown, processed, and marketed as "100 percent organic," "organic," or "made with organic ingredients" in the United States must do so under strict compliance with the National Organic Standard.

Organic Certification

If you're planning to market organic goats, with few exceptions your farm must be certified organic. Certification begins with official inspection, then the submission of an application and a detailed Organic System Plan. Once approved, a three-year conversion period is required to achieve full organic status. Facilities, animals, books, and records are reinspected on an annual basis.

Under the program, producers who market less than $5,000 of organic products per year aren't required to apply for organic certification. They must keep detailed records, however, to prove compliance with the National Foods Production Act and the national standards, and they're subject to National Organic Program audits.

One way or the other, to sell organic chevon, you must comply. If you call your goat meat organic and when challenged you can't prove it, you will be fined up to $10,000 per violation.

TO MEET CERTIFICATION STANDARDS

- Everything your goats eat must be certified organic: hay, grain or pellets, mineral mixes, milk replacer — everything. Bedding, if consumed, must also be certified organic.
- The pastures and woodlands where animals feed must be certified organic and maintained without the use of chemical fertilizers, pesticides, or herbicides. Pastures, hay meadows, and housing must be located far enough from neighboring properties to prevent chemical drift.
- Your goats must have access to outdoor areas, shade, shelter, space for exercise, fresh air, and direct sunlight. Shelters must be safely constructed, well ventilated, and designed to provide them with opportunity to engage in comfort behaviors as well as protection from temperature extremes.
- The use of antibiotics, genetically modified probiotics, and most other conventional veterinary medications is prohibited. Yet regula-

And compliance isn't a stroll in the park. The National Organic Standard outlines in great detail what you can or can't do with organic livestock. Before you proceed, read the standard at the National Organic Program website (see Resources, page 302) or download *NCAT's Organic Livestock Workbook: A Guide to Sustainable and Allowed Practices* from the ATTRA — National Sustainable Agriculture Information Service website.

Grass-Fed and Natural Chevon

A better option for producers eager to market goat meat to health-conscious buyers is to raise grass-fed chevon.

The American Grassfed Association defines grass-fed goats as those "that have eaten nothing but their mother's milk and fresh grass or grass-type hay from their birth until harvest." At this point, however, there are no USDA-mandated standards for grass-fed products, so the term is somewhat open to interpretation.

Consumers pay premium prices for grass-fed meat for a host of reasons. Pasture-raised animals are generally healthier than their feedlot-managed kin and their meat has definite health advantages. Various research studies

tions state that a producer may not "withhold medical treatment from a sick animal in an effort to preserve its organic status." Herbal medicines, homeopathic remedies, and most other alternative modalities are acceptable, as are most standard vaccines.

- Synthetic deworming agents are strictly forbidden, with the exception of the chemical wormer Ivermectin, which must never be administered to slaughter stock but is allowed for emergency treatment of breeding stock when all other strategies have failed. It can't be used to deworm breeding does during the last third of gestation or while nursing kids that will be marketed as organic.

Because certified organic feeds are costly and frequently unavailable, and because goats are so prone to internal parasites, it's nearly impossible to raise organic goats in most parts of the United States. Think carefully before undertaking organic production.

indicate the milk and meat of grass-fed ruminants are richer in vitamins A and E, beta-carotene, conjugated lineolic acid (CLA), and omega-3 fatty acids than those of their grain-fed peers.

Because goats thrive on a grass- and browse-centered diet, they're ideal candidates for grass-based farming. In fact, eliminating grain from their diet renders them less susceptible to enterotoxemia, urinary calculi, and other metabolic diseases. Grass-fed kids mature more slowly than grain-finished kids, however, so they can't go to market quite as soon. For the producer marketing from his or her farm, this doesn't pose a significant problem.

The USDA doesn't regulate the use of the word *natural* in describing food products. Depending on the producer, *natural* may mean near-organic, grass-fed, antibiotic-free, or a dozen other things, so when using the word to describe your product, spell out the specifics so your buyer understands what you intend to communicate.

Breeding Stock

Newcomers starting out in meat goats look to breeding-stock producers to provide them with healthy foundation animals, and market goat producers often come to them for replacement bucks and does. If producing these animals is your goal, your herd should consist of prolific, meaty does and bucks free of faults that affect meat production, and you'll need an economical means of maintaining them. They can be fullbloods or percentages of your favorite breed or productive crossbreeds, depending on the type of goat you want to sell. Unless you elect to market your stock through the auction barn — not the best place to fetch good prices — you'll need to advertise and promote your goats so buyers know what you have.

Show Goats

The most lucrative facet of the meat goat industry is the production of high-dollar show goats. Unfortunately, it's the hardest niche to master.

Each of the three Boer goat registries sanctions scores of shows held throughout the United States every year. The other registries don't do this, so if you want to breed show goats, you'll have to breed Boers.

The quality of the animals exhibited at these shows improves with every passing year. It's difficult (if not impossible) to "breed up" to show-quality Boer goats, so you'll have to invest in expensive bloodstock right from the start.

Also, the type of goat that wins at the big sanctioned Boer shows is rapidly changing. For better or worse, massive, short-legged, wrinkle-skinned South

SHOW YOUR GOATS

There isn't a school you can attend to learn how to show goats (although maybe someone should start one), so you'll have to teach yourself. Here's how.

- Find a mentor. There is no substitute for hands-on experience. If you know someone who shows, offer to help with preshow preparation and to act as a gofer at the shows. Show folks can always use another set of hands and legs, and you'll learn about showing right at the source.
- Become an adult 4-H volunteer. Ask to help whoever leads your county's meat goat project; attend training sessions with 4-H youngsters and learn right along with them. Fitting (preparing for show) and showing in sanctioned Boer show classes isn't precisely the same as 4-H market goat class protocols, but it's close enough that you'll learn a thing or two.
- Go to plenty of shows. Watch how the winners position and show their goats. Wander around behind the scenes and watch as exhibitors fit (prepare) their stock for show. If they aren't too rushed, ask questions. Goat owners are friendly folks; most will happily explain to you how and why things are done.

African–type Boers aren't winning like they used to and a leaner, tighter-skinned show ring Boer is emerging. Fads come and go; exhibitors have to stay on top of things in order to win, and breeders must produce that type of winning goat if they want to sell their stock.

The solution: Join the registries and study their periodicals to see which bloodlines are in the limelight. Attend some shows before you buy any stock and take note of what the top goats look like. Subscribe to Boer-oriented e-mail lists. Read and ask questions. Educate yourself before laying your money on the line for "show goats" of passé type and bloodlines.

Then ease into goats before buying your first expensive stock. While you don't want to lose *any* goats, it happens to everyone and it'll happen to you. Mastering the learning curve while raising $500 goats is less financially devastating than losing a $3,000 ennoblement-pointed doe, and you can practice your fitting and showing skills by taking your $500-goats to county and regional fairs.

To make a name as a producer of quality show goats, you *must* show. If you don't care to do it yourself, you can hire a fitter (a professional who prepares and shows goats for other people) to do it for you, but you need show-world exposure to demand top dollar for your goats.

Finally, you must actively promote your business. Advertising in breed periodicals is a good first step, but pursue the promotion strategies listed in chapter 15 as well. The more recognizable your name, the more confidence buyers will have in your product: your goats. And to keep that good name, stand behind the goats you sell!

An Unexplored Market: Working Goats

You won't make money hand over fist selling working goat prospects, but for goat ranchers looking to diversify, it's an additional market for excess male kids.

Working goats (also called recreational goats) are trained to pull carts and wagons or they're taught to pack goods in the manner of scaled-down pack-horses. Because male goats grow larger than does (but most recreational owners prefer not to deal with the aggression and odors of bucks), working goats are generally wethers.

When picking a kid to raise as a working prospect, choose a young-ster with good-size parents. An adult working wether should measure at least 34 inches tall at the withers and weigh in the neighborhood of 200 pounds. He should be wide and deep through the chest and show plenty of width across his hip and shoulders, and his legs should be in proportion to his body. He'll need plenty of leg bone and large hooves to remain sound. His elbows should lie close to his body, even when he's moving, and if his forearms are longer than his cannon bones, he'll take long strides. Packgoat kids have an added requirement: a nice level

Packgoats make camping easier and much more fun.

top line. (To carry a pack most efficiently, a packgoat's withers and croup should be the same distance from the ground.)

Attitude is important too. Packgoat fanciers use the term "gung-ho goat" for the sort of animal that makes a first-class working goat. Working goats should be friendly, curious, and sensible. If a kid resists (or worse, resents) handling, he's not going to make an efficient working wether.

Packgoats are popular with outdoor adventurers for a number of very good reasons.

- They're the perfect solution for anyone seeking an outdoor experience but who doesn't want to (or can't) shoulder a heavy backpack on the trail. Hunters favor them for packing out meat and they're ideal for older campers or families with young children who couldn't otherwise tote enough camping gear by themselves.
- Their smaller stature and ease of handling make goats more user-friendly than packhorses, donkeys, llamas, and yaks. They can be hauled in a goat tote or pickup topper instead of a trailer; no ramps or loading facilities are necessary because they readily hop in or out of a pickup bed all by themselves.
- Goats want to stay close to their handlers. Unless trail rules dictate otherwise, properly bonded packgoats needn't be led on the trail or tied in camp at nightfall. Loose goats browse along the trail, so you needn't bring along any goat feed.
- Although packgoats can't carry much weight until they're full grown, a three-year-old adult packgoat can tote 20 to 25 percent of its own body weight — roughly 50 pounds of food and gear. Even a kid can carry a training pack on the trail.

Gung-ho kids can pack a few sandwiches in a Pygmy-size training pack by the time they're five or six months old.

- Goats are long-lived. Barring illness or accident, most packgoats have a working life of 10 to 12 years.
- Packgoats cost comparatively less to buy or raise and maintain than other pack animals. It costs less to outfit packgoats too; a quality pack saddle and top-end panniers cost in the neighborhood of $300.

Most packing activity takes place in the rugged western mountain states, where dairy wethers' longer legs and extreme agility make them the packgoat of choice. Packgoat fever is spreading throughout America, however, and barring areas where sustained rock climbing is required, meat goats' considerably greater carrying capacity than that of dairy breeds makes them shine.

If you want to market packgoat kids, plan to bottle-feed them. Some enthusiasts prefer to buy two- or three-day-old bottle kids to raise themselves; others will pay a premium for weaned bottle-fed kids. They don't want wild,

CLICKER-TRAINING GOATS

Clicker-training, also known as *operant conditioning,* is the fun and stress-free mode of training used by aquarium and show-biz trainers, zookeepers, and hundreds of thousands of horse and dog owners to teach their charges tasks as diverse as fetching a can of soda, leaping through fiery hoops, and helping rescue personnel locate victims of disasters.

When animal students perform a desired action, the click of the trainer's handheld clicker says "Well done!" The click is quickly followed by a tidbit of a favorite food. Goats, being the intelligent (and food-loving) creatures they are, are ideal candidates for this gentle, effective mode of training.

Clicker training is easy, builds a strong bond between student and handler, and requires little in the way of specialized equipment. Because training methods pioneered for other species work for goats, a 50-cent clicker and a book on clicker-training horses or dogs will do nicely for starters. *Clicking with Your Dog: Step-by-Step in Pictures,* by Peggy Tillman (Waltham, Mass.: Sunshine Books, 2000) is the clearest introduction to clicking we've ever seen.

dam-raised youngsters: it's important that packgoats bond with people right from the start.

Or you can ask more money for started kids in the 6- to 12-month-old bracket. They should have good manners; be easy to catch and handle; and be trained to tie, lead, and stand. You'll want to take them on a dozen or so cross-country walks (sans packs, collars, and leashes) to be sure that they'll readily follow a human along the trail.

Young packgoats should be fully socialized by whatever age you intend to sell them. They must be tame, friendly and mannerly, meaning they aren't allowed to (and know they can't) jump up on, mob, or otherwise maul a human.

If you have a bountiful supply of browse or pasture available, weaned future packgoat wethers can be raised to near maturity on forage alone. Their hooves must be kept in absolute tip-top shape and they should be vaccinated according to your meat herd's health maintenance protocols (they'll need CD/T inoculations at the barest minimum). Most packgoat producers expect purchases to be tested for CAE and sometimes CL, so you'll need to factor that cost into pricing your stock. Those who sell kids to raise as packgoats really needn't spend a lot of time or money preparing youngsters for a career in packing, and the folks who buy them generally give them stellar, long-term homes.

GOATSTOCK, THE NATIONAL PACKGOAT RENDEZVOUS

Once each summer the North American Packgoat Association sanctions Goatstock, the National Packgoat Rendezvous. Always held in one of the scenic western states, these three-day goat gatherings provide great campsites, delicious outdoor food, hiking opportunities, educational seminars, and plenty of camaraderie for all admirers of working goats (harness goats and their humans are encouraged to come to Goatstock too). To learn more about this year's Goatstock as well as upcoming regional packgoat rendezvous (there are several every year), and to read about Goatstock's past, visit the North American Packgoat Association website (see Resources).

Harness Goats

In other times, and even now in parts of the world, goats have earned their keep by pulling all kinds and sizes of carts and wagons, sometimes alone and sometimes working in a team of many goats. As recently as the 1970s, the legendary traveling preacher Ches McCartney ("the Goat Man"; read more at http://en.wikipedia.org/wiki/Ches_McCartney) traveled the United States in a covered wagon drawn by nine or ten goats. Today, most goat-driving enthusiasts drive single and team hitches of goats strictly for fun.

A sturdy, full-grown, meat breed wether can easily pull an adult and a child or two when hitched to a well-balanced cart or a four-wheel wagon. Teams of two goats are generally hitched to wagons and can pull as much weight as a stocky pony or small horse.

Like packgoat prospects, the kids you sell as future harness goats should be friendly and outgoing, and while they're still quite small they should be taught to tie, lead, and stand still. You'll also want to encourage them to walk on lead in front of you, rather than at your side, because a goat's natural impulse is to follow.

It's also a viable option to sell started or fully trained driving goats. Little has been written about training goats to harness, but it's much like training a driving horse or pony, with a few minor twists.

A full-grown goat can easily pull a well-balanced cart. While wethers are the driving goats of choice, does and bucks can work in a harness too.

- Equines want to lead; goats prefer to follow. Be prepared to spend a great deal of time on the ground driving or enlist the aid of an assistant to lead your goat for a while until it catches on.
- Goats are more curious than horses and ponies, so until they're fully seasoned, allow them to carefully examine equipment each time you expect them to wear it. Take caprine nature into consideration when training a goat to drive. Don't lose your temper. Be patient — always.
- Food motivates goats far more than it motivates horses. Reward good behavior. Consider clicker training (see page 236); it's the best all-around way to train a working goat.
- Most ready-made goat harnesses utilize halter-style bridles rather than relying on horse-style bits for control. It's important to train harness goats to respond to all voice commands but especially to stop reliably when they're told *Whoa*.
- It can be very difficult to find ready-made goat-size harnesses for sale and even harder to locate quality goat carts or wagons. You can find both on eBay or buy them from companies like Caprine Supply and Hoegger Goat Supply. And it's reasonably easy to fashion your own harness out of nylon strapping. For plans, visit www.dreamgoatannie.com and click on "Goats."

This goat wears a halter-style bridle.

THE TRAVELING PHOTOGRAPHER AND HIS GOAT

Have you ever admired a vintage photograph of a smiling child seated in a scaled-down Studebaker wagon or spiffy cart, holding the reins of a handsomely harnessed long-horned goat? If so, you've admired the work of one of hundreds of photographers who traveled the United States between 1880 and 1950 taking pictures of children in an era when few families owned a camera. They posed their wee subjects with animal props, sometimes astride a pretty pony or a solemn donkey, but most often in a cart or wagon pulled by a goat.

Producers Profile

Pat and Clark Cotten, Bending Tree Ranch
Damascus, Arkansas

Pat and Clark Cotten's Bending Tree Ranch is nestled at the base of the Ozark Mountains in north-central Arkansas's rugged goat country. Their goat business is one of only two Onion Creek Ranch satellite operations in the United States. The Cottens raise fullblood Boer goats as well as a selection of all of the breeds produced by author and goat producer Suzanne Gasparotto on her Lohn, Texas, ranch. Pat is an instructor at Onion Creek Ranch's famous GoatCamp, a weeklong, hands-on training program for commercial goat breeders held at the ranch every October, as well a member of the board of directors for GoatCamp and director of the Onion Creek Ranch Satellite Ranch Program.

Q *Pat, how did you get involved with meat goats?*

A We got started in meat goats as a 4-H project for my son back in 1997. His first goat was a Boer/Nubian cross and it needed a buddy, so we bought a Myotonic/Boer cross doeling from my herding instructor to be its companion. She turned out to be the meatier, more heavily muscled of the two. So we started researching the Myotonic breed, and soon after that we bought some full Myotonic stock to add to our herd of Boers and Boer crosses.

Q *What facet of the meat goat industry do you serve?*

A We breed and raise competitive, high-quality, market show goats, but our main business is raising top-quality replacement breeding stock for commercial producers. We focus on mothering abilities, kidding ease, good feed conversion, growthy kids [those that mature quickly and well], and hardy stock, so our breeds of choice are Myotonics, Tennessee Meat Goats, TexMasters, Boers, and crosses of these breeds. We choose these breeds because of the higher meat-to-bone ratio of the Myotonic and Myotonic crosses, their better feed conversion, their excellent mothering instincts, and the overall hardiness of these animals.

Q *Do you feel there's room in the meat goat industry for new producers?*

A I get calls several times a week from folks just getting into the industry or researching the market before jumping into the business. I can't keep enough breeding does available to meet the demand for quality breeding stock. There is plenty of room for new producers.

Q *Do you have any advice for prospective goat raisers?*

A Raising meat goats isn't a "get-rich-quick" business. To succeed, you need to know the market demand in your area and build your herd around that demand. What works for somebody in another part of the country may not work where you live. Know who the buyers are and what they require as far as size and age, when the holiday demand is, and so on.

Have the right kind of fencing *before* you buy your goats. Make sure you have some kind of livestock guardian *before* you have a problem with predators. We run Anatolian shepherds with our goats and will never be without them. They are invaluable to our operation. We've never lost a goat to any predator and we live in an area heavily populated with coyotes.

And do your market research *before* making your first purchase. Attend seminars offered in your area. Check with your state Extension service. Look for goat-related organizations in your state. GoatCamp is an excellent educational resource for commercial producers. It's a one-of-a-kind camp run strictly to educate goat producers, whether they're thinking of starting a goat operation or have been raising goats for years.

Educate yourself! I simply can't say this strongly enough.

15

Promoting Your Goat Business

YOUR FLEDGLING GOAT BUSINESS WILL NEED promotion in order to thrive. Of course, first you'll have to settle on a catchy name for your enterprise. After you've determined a name, in this day and age a primary source for advertisement is the Internet — but don't underestimate the power of more traditional avenues of advertisement such as industry advertising, business cards, and flyers and brochures. Think of all of these strategies working together to reach the greatest number of potential customers and contacts for your growing goat business.

Choosing a Business Name

Before creating a website or other advertising pieces or mapping out a business plan, you'll need to choose a business name. Make it unique. Make certain people can remember, spell, and pronounce it. If you register goats, don't choose a name another producer within your breed is already using. (To determine if a name you've thought of is taken, contact appropriate registries and ask.) Finally, to prevent unpleasant litigation, it never hurts to find out if your chosen name is trademarked; if it isn't, consider having it trademarked.

Websites Sell Goats

If you produce show and breeding stock, you definitely need a website. You also need a site if you market slaughter goats from your farm and you want folks to be able to find you. In today's hectic world, the Internet is a customer's quickest way to locate the goods he or she desires. Make it easy for customers to find you and they'll buy.

What, specifically, can a website do for your goat enterprise?

- You can show your goats and present your message to an unlimited number of people from all parts of the world, without leaving your home or paying for a single pricey ad.
- You can give potential buyers more information than you could in an ad or typical brochure and you can change it whenever you want, from home, 24 hours a day.
- You'll automatically target an audience interested in goats or in goat meat. (Only those interested in goats will visit your site.)
- On the Internet your well-run backyard meat goat business can seem as impressive as those of your biggest competitors. It might even seem better if you're an able photographer and skilled web designer (or if you hire a designer with great skills).
- You can display all of your merchandise (live goats, goat meat products, goat milk soap, meat goat T-shirts, Boer goat halters, plush toy goats) and tell about each item rather than highlighting just a few choice items.

Maintaining an Internet presence is a remarkably cost-effective way to market and promote your goats and goat-related products, especially if you do the work yourself. A comprehensive, do-it-yourself website costs about an hour or two a week in time and $200 or $300 per year to build and maintain, although you'll have to master certain skills to do it well. If you're a technophobe, you can go the alternate route and hire a developer to create and maintain your site. Whichever avenue you choose, make sure your site is well constructed and attractive.

Work with a Pro

Before hiring a webmaster to build your website, surf the Internet and pick out other sites you like. Contact whoever built them. This information is generally found at the bottom of a website's home page. If it isn't, e-mail the site's owner and ask.

WEBSITE WISDOM, THE MAC GOATS WAY

Claudia Marcus-Gurn and her husband, Matt, of MAC Goats raise fine show- and breeding-quality fullblood Boer goats on their ranch in the Missouri Ozarks. The offspring of their herd sires — Downen R33 "Hoss" Ennobled and EGGS P653 "Jubilation" Ennobled — are ever-increasing forces to be reckoned with in the show ring. MAC Goats is perhaps best known, however, for its information-packed website, a mecca for meat goat owners of every persuasion.

Claudia, who single-handedly maintains the site, offers these thoughts on goat business websites: "A website is a very important means of displaying your product if you're breeding show and breeding stock. For us, it's vital. A website keeps breeders in touch with customers distributed throughout the United States — and overseas, too, for that matter. When we moved here from California, our customers 'moved' with us because of the web site. Besides displaying our product, our website also performs a service for those interested in the Boer goat; education is its number one priority. Right from the beginning, as I've learned new things, I've shared them on our website because I know others also want that information. I think that by educating potential customers, we help them make good choices no matter who they decide to buy from. Improving the breed and encouraging a successful operation for us and others is what it's all about for me, and our website helps contribute to that end."

MAC Goats' website is http://members.psyber.com/macgoats, and Claudia and Matt's bricks-and-mortar contact information is HCR 89 Box 409, Winona, MO 65588, 417-778-1904.

Collect a list of potential designers, keeping in mind that web designers' charges can run the gamut from pizzas and beer to fuel your computer-geek son-in-law while he builds your site to $80 an hour and up for certain professionals. (Be aware that your son-in-law might build a better website!)

Ask each prospective designer how much bang you'll get for your buck: How many pages does his or her base fee get you — and how many pictures per page? What else is included in the base price and what features, if any, can be had at extra cost? How many updates are included? How much are additional updates beyond what's included? Will the developer work with you and your input or insist on doing things his or her way (some designers might be tied to certain bells and whistles, but technology doesn't sell goats). How long will it take to get your working website online? Will the developer list your site with major search engines or is that your job? Does he or she offer a satisfaction guarantee? Ask for URLs (website addresses) to other sites the designer has built and then e-mail their owners. Are they happy with the work? If not, why not?

Do It Yourself

Unless you're computer-illiterate and tremendously strapped for time, building your own site is the best way to go. You know precisely what you want, and you're much more likely to get it when you do it yourself.

Check with schools and libraries in your locale; many offer free or inexpensive weekend or evening computer classes, including classes on site building and Internet commerce. In addition, county Extension services hold Internet farm marketing seminars and the Small Business Administration (www.sba.org) can also help you to get started with building your site. You might also choose one of the scores of up-to-date, easy-to-understand books designed to guide beginners through the intricacies of building a website. It costs less to bypass professional designers and build your own site; updates are easy and free; and the satisfaction of designing your own site is priceless.

Are you game? You'll need the following.

A Plan

Websites sell something or distribute information; the best of them do both. Decide what yours will do and be. Plan each page on paper, based on the principles below. Remember: It's easier to build from a blueprint than to backtrack and fix things later.

Web-Page Authoring Software

HTML is the programming language of the World Wide Web. Go to your computer and navigate to virtually any website, then choose "View" in your browser's tool bar and "Source" from the drop-down menu. This will allow you to view the page's HTML — a stream of letters and codes and symbols. Though it looks complicated, you won't have to learn to write it if you use software for building web pages. Such software is a must-have for novice site

WEB-BUILDING ASSISTANCE: FIND IT ONLINE

- If you need to download free or low-cost web-building software or require any other form of website construction assistance, visit one of these web-building resource sites. They've been helping newbies for a long, long time and they're free (or inexpensive):
- Big Nose Bird, www.bignosebird.com
- Web Monkey Web Developer's Resource, www.webmonkey.com
- Build Your Own website, www.groupasystems.com

builders, and you needn't break the bank to buy the kind you need. Big-name programs such as Dreamweaver and FrontPage can be used to build wonderful, high-tech websites, but for selling goats, these might offer more than you want or need. For your purposes, free or inexpensive programs such as WebPlus, WebMaster, WebStudio, and Ewisoft Website Builder work just fine. I build and maintain our half-dozen websites using Claris Home Page 3.0 software, written in 1997, which I purchased brand spanking new (though outdated) on eBay for $12. Plain-Jane but fully functional building programs can also be downloaded direct from the web.

A Great Domain Name

Your domain name is the "who" in your website's Internet address. Register it right now. Catchy names are purchased and taken out of circulation every day, so don't lose yours because you've waited too long. Use your business name if you possibly can. If your business name is already taken, try customizing it by adding hyphens (Ozark-Thunder), underscores (Ozark_Thunder), numbers (OzarkThunder1), or additional descriptive words (OzarkThunderBoers, OzarkThunderBoerGoats, OzarkThunderRanch). If none of these strategies works (that is, all of them produce names that are unavailable), brainstorm another easy-to-remember, goatish name. These catchy names may still be up for grabs: WorldsBestGoats, TastyMeatGoats, and BestGoatMeatOnEarth — and you can probably think up an even better name! If your first-choice domain name is taken (let's say it's FantasticBoerGoats.com), you still might be able to register it as FantasticBoerGoats.net, FantasticBoerGoats.biz, or FantasticBoerGoats.us. To determine if your favorite choices are already taken, type "domain name check" in the search box at your favorite search engine and use any of thousands of free, online domain name searches to find out for sure. A tip: Domain names typed into a browser bar aren't case-sensitive — that is, don't worry about typing capital letters at the beginnings of words. For instance, www.ozarkthunder.com works the same as www.OzarkThunder.com. But when adding your URL to e-mail signature lines, directories, or printed promotional items such as business cards and brochures, capitalizing helps folks recognize your name.

Before you register your domain with an independent registration service, check to see if your hosting server includes domain name registration in its monthly service charge. If it does, use this service and save yourself some money.

A Hosting Service

Here's a cardinal rule of website construction: Don't use freebie hosting services to host your business website. People hate the advertising banners that are part and parcel of sites on freebie services. Some freebie services are worse than others. I won't follow links to sites on several of the major free hosting servers because they're notoriously slow to load, treat me to pop-ups on every page, and are prone to freezing Macs and older PCs. Some of them also automatically load spy- and ad-ware on visitors' computers. Most of them are best avoided.

In addition, most freebie services equate with cumbersome URLs. Which will your customer remember: www.OzarkThunder.com or www.dontusethis service.com/~123homer/ozarkthunder? Another reason to avoid these hosting services: Most major search engines' web crawlers often skip over freebie service sites, giving priority to sites with bona fide domain names. Finally, freebie services sometimes abruptly switch to paying status. If this happens, you'll have to shell out instant dollars or see your site go down the drain. Is it worth it?

Instead, choose a service you can live with, taking into account the following considerations:

- **How much bandwidth will you get for your money?** (Bandwidth is the number of bytes [usually expressed in gigabytes, or GB] consumed by your customers' visits each month.) More is always better. Remember that when you run through your monthly allotment, your site goes down until the next billing period begins.

- **How fast does it connect to the Internet?** Connection speed varies; some services are infinitely slower than others.
- **What does its support system consist of?** Is there well-written online documentation you can refer to when you have questions? Can you contact support staff by e-mail or phone? How fast are e-mails answered? Are tech-support personnel willing to patiently walk new designers through problems, using everyday language instead of technical jargon?

An easy-to-use hosting service is a jewel beyond price. For your first foray into site building, you might do best to choose a major hosting player such as Yahoo Web Hosting (http://smallbusiness.yahoo.com/webhosting). Yahoo Web Hosting's online help feature is written in a way that beginners can understand, its web-building tools are superb, and its tech support is as helpful as can be. We started hosting our business with this one for its ease of use, and 10 years later, it still hosts our sites.

Website Design for Goat Entrepreneurs

Whether you build your own website or a developer does it for you, you'll want it to be eye-catching, effective, and user-friendly. Most websites are not.

According to the USDA Marketing Service publication "How to Direct-Market Products on the Internet," these are the factors most likely to influence repeat visitors to a business website: seventy-five percent of visitors expect high-quality content, 66 percent praise ease of use, 58 percent won't revisit slow-loading web pages, and 54 percent avoid dated sites. Only 12 percent of visitors go back to a site to view cutting-edge technology.

As Ellie Winslow, author of *Living Beyond the Sidewalk* and *Making Money With Goats*, points out, "Over the last decade websites in general have gotten more glitzy. If the website is for business, don't distract your visitor and don't wear him or her out with all the moving icons, streaming banners and other technically advanced stuff that doesn't actually promote your marketing goals."

For selling purposes, websites should present products and facts in a plain but attractive, easy-to-use manner. Keep in mind the following important considerations.

Loading Times

According to recent studies, the average Internet user spends nine minutes a day — that's almost 55 hours per year — waiting for web pages to load. For each person who patiently waits six minutes for your picture-rich and gizmo-laden web page to appear, another will allow 30 seconds, then leave.

Many web developers assume every Internet user has lickety-split, quick, direct satellite link (DSL) service — but most don't! In rural areas (precisely the regions where you want to market goats), dial-up service is still the norm, often via antiquated phone lines.

Further, because many goat people hang on to dated computer equipment, preferring to use their dollars to buy goats instead of faster computers, it's vitally important to design pages that load quickly and grab visitors' attention before they leave to peruse another goat seller's site!

Remember KISS (Keep It Simple, Sell). Web designers recommend that a web page's elements add up to single-page downloads no greater than 180 kilobytes (KB). This precludes huge pictures and technical froufrou, so you have to plan your layouts very carefully.

If you love bells and whistles, build a dual website linked through a common home page. Let visitors click links to enter a no-frills version or one resplendent with music, glitz, and animation. Add a hit counter to each; you'll be surprised which one most visitors choose.

Images

Resize large photos and other graphics so they're attractive yet will download quickly. Software such as Adobe PhotoDeluxe and Adobe Photoshop, or any of the free and inexpensive photo-editing software available online to download, will help you accomplish this. Resolutions close to 72 pixels per inch are ideal for website applications.

Plan your pages carefully to eliminate photo glut on any one page. Start with a spare, uncluttered introductory page for each product (such as an individual goat) and link to additional pages of photos and information.

The cardinal rule for using photos on your website is to choose good ones! See appendix B, Photograph Your Goats, for information on taking good goat pictures. If you're short on patience or persistence, hire a professional to shoot clear, well-composed photos of your goats.

A word about photos and graphics: Don't assume everything online is up for grabs. While you should never "borrow" an image or graphic from another site without previously clearing it with that site's webmaster, many site owners are happy to share if you call or e-mail and ask.

The Rest of the Story

Your website is in essence your goat business retail store, so it behooves you to build a good one. We can't cover all there is to know about web design in this short chapter, but these additional tips should help you get started.

- **Don't assume your visitor is Internet-savvy.** Many people aren't. If they don't understand mouse-over images and pull-down menus, they'll miss a lot of your site's content, so place information and important images where people can easily find them.
- **A visitor probably won't be back** if your site crashes his computer — another good reason to steer clear of animations, music, and other website gewgaws.
- **Don't use frames!** Sites with frames are cluttered and confusing, and they rarely print well. They also load poorly on older computers and small computer monitors, overfilling the screen and sometimes blocking access to your site's best features. If you must use frames, don't place contact information near the bottom because that's the part that generally doesn't fit on such screens. A best bet for frame aficionados is to design two sites linked to your home page, one with frames and one without, which allows visitors to choose the version they prefer.
- **Choose your fonts with care.** Make certain they're large enough for a viewer to read them easily on both Macs and PCs and in all of the standard browsers. Avoid nonstandard fonts; if in doubt, use Ariel or Times New Roman.
- **Don't incorporate patterned backgrounds;** they tire visitors' eyes and copy can get lost in the morass. Strive for clear contrast between font and background colors. Light-colored copy against a dark background looks great online but doesn't print well in older versions of certain browsers.
- **Your home page should make clear what your site is about** and what it has to offer your visitors. It should be welcoming and uncluttered and should load in a snap. You're allowed only one first impression: make it count.
- **Include a site map on your home page** and provide a link to it from each of your other pages. Visitors stay longer when they know what's there to see.

- **Compose headlines and page titles that help make sense of what's on each page.** Search engines frequently link to internal web pages. When this happens, without some pertinent description on the page, a visitor feels lost.
- **Place your name, contact information, and logo on every page** and link the logo back to your site map or home page, because not every visitor enters your site via the home page. Don't omit your physical address. Connecticut buyers might not be interested in goats from Washington, but if you're only a county or two away from Joe Goatbuyer, he's much more likely to call and arrange a visit.
- **Keep URLs simple.** Visitors sometimes type them to access internal web pages. Keep these simple. Use all lowercase characters and avoid special characters such as underscores (_) and tildes (~) whenever you can.
- **Write good copy.** Buy a book, visit a website, or take a class on effective verbal sales strategies before you put up your site. One helpful book is *Living Beyond the Sidewalk: How to Thrive There,* by Ellie Winslow.
- **Triple-check all copy** for typos and misspellings. Don't make nitpickers grit their teeth.
- **Add educational content** — the more, the better. When goat owners visit educational sites such as those maintained by Fias Co Farm, MAC Goats, Bar None Meat Goats, and Jack and Anita Mauldin, they often stay to see what's for sale. They'll be back next time they need information and they'll tell their friends. This is prime advertising!
- **Don't wait for search engine web crawlers to find your site.** Go directly to the major search engines and sign it up.
- **Don't let your site become dated.** Tweak your sales lists, upload pictures of those cute new kids, or add a new educational item — keep your content fresh and bright. And spend time each week checking links, especially those that lead to outside sources. One hour a week spent on website maintenance pays big dividends in customer approval.
- **If someone e-mails via your website, respond as soon as you can.** When marketing via e-mail, punctuality counts. According to a study reported in "How to Direct Market Farm Products on the Internet," by Jennifer-Claire V. Klotz (USDA Agricultural Marketing Service, 2002), 40 percent of the top e-commerce websites took longer than five days to respond to e-mail, never replied, or simply weren't accessible via e-mail. People want results right away. Give it to them and sell those goats!

- **Aggressively promote your site in every way you can.** People won't visit unless they know about it. Add it to your e-mail signature line so it appears on every e-mail you send. Use it on your business cards, brochures, and road signs. Letter it on your truck and have it emblazoned on the T-shirts you wear to the store. Swap links with other goat producers. As a goat entrepreneur, your website is your most important marketing tool. Make it a great one and consider your time (and money) well spent.

Industry Advertising for Pennies

Most new goat entrepreneurs assume ads in major livestock publications cost the world, but this isn't necessarily so. Marketing statistics indicate sellers get a bigger response from small ads placed often than they do from occasional splashy, full-color, full-page ads.

So consider advertising in industry publications like *The Goat Rancher* and breed journals such as *The Boer Goat*. Both offer inexpensive options to fit every budget. You needn't break the bank to advertise nationally when you utilize bargain-savvy special rates.

Make Business Cards Your Ace in the Hole

Dollar for dollar, the most effective advertising in the world is a top-notch business card. This seller is inexpensive and exquisitely portable, and if it's designed right, people tend to keep it for a very long time. There are scores of ways to promote your business with good cards. Never leave home without them!

- **Tack cards to every bulletin board you encounter,** whether at the supermarket, Laundromat, feed store, or any community posting place. Use sturdy push pins so you can stack cards, which encourages interested parties to take one along. Check your posting places often to restock cards.
- **Purchase inexpensive business-card holders and ask businesses to display your cards near their cash registers.** Target veterinarians and farm stores in your area, but place them in other local businesses too.
- **Tuck a business card into every piece of mail you send out** — personal correspondence (Great-aunt Tilly may know someone who'd like a goat), invoices, even bank payments and the electric bill. Some will likely be tossed out, but others won't.
- **Craft your own gift tags** — on the back of your business cards, of course.

- **Place cards in library books to use as bookmarks** or insert cards in goat books and magazines sold through farm and book stores (though you should always ask the management before you use this avenue).
- **Hand them to people you meet.** Ask them to take several cards and pass along the extras to their family and friends.
- **Buy a conference-style name tag holder, insert your card, and wear it on your lapel;** it's certain to stimulate conversation!
- **Scan and add it to your e-mail** as an attachment.
- **Use your card as a camera-ready ad** for publications; it'll save you money on setup fees.

Build a Better Business Card

The only thing worse than not using business cards is distributing bad ones. Soiled, poorly designed, cheaply made cards affixed to bulletin boards or stacked on the counter at the feed store create an impression of your business that you definitely want to avoid. Quality cards, on the other hand, tell an entirely different tale. Make this first impression a good one!

Though it's tempting to settle for freebie cards you can order online, don't go this route. The advertising on the back of these cards detracts from your image. *Cheap* and *shoddy* aren't words you want associated with your goat operation, so go the extra mile and pay for your cards.

If you spring for quality cardstock (not perforated punch-outs) and you understand design, you can probably print your own. Many word-processing and most desktop-publishing programs offer create-a-card capability or use downloadable templates available from suppliers such as PaperDirect. If you can't do a professional-looking job, however, you're better off ordering business cards printed by the pros.

MORE IS LESS

Don't buy small quantities of business cards — the more you order (usually in multiples of 500), the less the cost per card. One thousand business cards is a working minimum if you plan to aggressively promote your business using them.

LONELY GOATHERD FARM

Specializing in packgoat tours and breeding Boers

P. O. Box 202
Okaogam, WA 98840
Phone: 000-000-0000
www.lonelygoatherd.com

An attractive, well-designed business card is the goat entrepreneur's least-expensive advertising tool.

Before designing your card or hiring a professional to do it for you, collect business cards you especially like. They needn't be goat-related. What you're looking for are elements that engage your eye and can be incorporated into your own card's design.

- **Include contact information.** How do you want people to find you? Consider including your name, farm name, mailing address, e-mail address, phone and fax numbers, and addresses to all pertinent web pages.
- **Keep it simple.** Avoid fancy, hard-to-read fonts and use one (or at most two) font families per card.
- **Your name (or farm name) should be the largest text element** on your card. Make sure additional text is easily readable.
- **Don't use nontext elements unless they're good ones.** Consider having a custom logo designed for your business. If you have no logo, good sources of infrequently used graphics are the books in Dover's clipart series. Many of these books include easy-to-use CD-ROMs. Images from other sources will need to be scanned and resized for use on your card.
- **A bad photo is infinitely worse than no photo** and its resolution needs to be spot on. Consult appendix B for goat photography tips or hire a pro to shoot a top-flight photo for your card.
- **Find a way to make your card stand out.** Textured cardstock and interesting cut-outs (use a scrapbooking paper punch) help make a card unique.
- **Choose cardstock and font colors that contrast with and yet complement one another.** Color adds interest, but a little color goes a long way.
- **Stick to standard size:** 3.5 inches by 2 inches. Because other sizes don't fit ready-made card holders, they're likely to be dumped in the trash.

- **People will save your card if you print something useful on the back.** Be creative. For instance, text like this is catchy: "New to Boers? To learn more, visit these online resources," followed by the web addresses of your favorite information sites. You might print a favorite goat BBQ recipe or a simple formula for calculating kidding dates.
- **Spell-check your finished card before you print or send it to the printer,** and then proofread it by eye several times. Nothing says "amateur" more than advertising replete with a bunch of misspelled words.
- **Create color interest at minimal cost by carefully rubber-stamping a small design element** on basic black-and-white business cards using colored or metallic ink.
- **Keep your cards current.** Don't scratch through or white-out text and write in corrections — instead, order new cards with any updated information.

Business Card Etiquette

Practice handing out your card, especially if you tend to be shy. Don't apologize or seem apologetic about presenting your card; smile, make eye contact, and memorize a simple, catchy line to say. Hand out your card faceup, with pride. This tells recipients you're proud of your farm and your goats.

If someone hands you his or her card in return, don't immediately stuff it in your wallet or pocket. Examine it. Hold it in your hand for a few moments as you converse. When you treat someone's card with respect, that person will likely extend the courtesy to you.

Finally, don't distribute bent, soiled, or tattered-looking cards. These will tend to reflect poorly on your business. Take care of your business cards. Keep them looking crisp and new in an attractive business-card holder or tucked away in a safe pocket in your wallet.

Create Your Own Flyers and Brochures

A flyer (sometimes called a circular) is usually printed on one side of standard 8½-inch by 11-inch paper and is designed to be tacked to bulletin boards, taped to the inside of windows, and distributed as handouts. They're an ideal venue to announce an upcoming production sale, spotlight the stud buck you stand to outside does, or simply indicate you have goats (or goat meat) to sell.

A brochure (also called a pamphlet) incorporates more-detailed information. It can be as simple as an 8½-inch by 11-inch or 8½-inch by 14-inch single-sheet trifold or as detailed as a multipage, spine-stapled booklet. An

WILDERNESS ADVENTURES CUSTOM-TAILORED FOR YOU!

- Trained guides, packgoats, and all camping gear provided
- Gourmet meals and trailside snacks (vegan or vegetarian on request)
- Day, overnight, three-day, and weeklong trips
- Multiple destinations
- Adventures for beginning goat campers are our forte

QUALITY PACKGOATS FOR SALE TO APPROVED HOMES

- Registered Boers, Saanens, and Boer-Saanen crosses
- Certified CAE- and CL-free herd
- Socialized, bottle-fed kids
- Trail-indoctrinated, pre-training-age goatlings
- Well-started one- and two-year-old packgoats
- Fully trained, seasoned packgoats occasionally offered
  (there is a waiting list for these outstanding goats)

Lonely Goatherd Farm
P. O. Box 202
Okaogam, WA 98840
Phone: 000-000-0000
www.lonelygoatherd.com
proprietor@lonelygoatherd.com

Simple flyers make outstanding handouts and bulletin-board posters.

effective brochure makes the reader want to open it and read what's written inside. It must be visually appealing and well written and should satisfy a need or send readers a specific message.

Both flyers and brochures have a place in your advertising arsenal. While you can certainly have these items custom printed (and if you don't care to master home publishing, you really should), in this age of desktop-publishing software, high-end printers, and quick-print shops, you can create effective flyers and brochures right at home. Keep these tips in mind.

- **Determine what you want your goat business flyer or brochure to achieve** and focus on this goal. If you raise Kiko goats, Cheviot sheep, and Longhorn cattle, don't try to market them all through a single brochure.
- **Choose headlines that draw attention to specific points.** Avoid artsy fonts that are difficult to read. Basic sans-serif fonts such as Helvetica, Arial, and Geneva work best. (Serif fonts have small strokes at the ends of the strokes in each letter; sans-serif fonts don't.)
- **Text should be written in a standard serif font** such as Times, Times New Roman, or New Century Schoolbook. Don't use script fonts of any kind.
- **Avoid using all capital letters in both headlines and text;** they're both distracting and difficult to read.
- **Don't crowd too much onto a single page.** Increased space between lines of text improves readability. Double-space most text (or use 1½ spaces between lines, at a minimum). Adequate margins on all sides of each page add elegance and increase readability too.
- **Keep sentences short and to the point.** Strive for simplicity and clarity in the copy you write. If you're unsure about what you've written, ask someone unfamiliar with meat goats to proofread it before it goes to print.
- **Always, always run a spell-checker over everything you've written,** but after you do, proofread the text several times to catch errors the program missed.

13 Additional Easy, Low-Cost Promotion Strategies

1. Invest in quality truck and trailer lettering. Its design should incorporate your phone number, e-mail, and website address. You can also turn non-business vehicles into rolling billboards for your farm by affixing to them custom-designed magnetic signs.
2. Order T-shirts, jackets, and hats printed or embroidered with your logo and farm name and wear them everywhere you go. Buy extras for your customers and friends.
3. Choose custom checks, invoices, and other business forms printed with your logo and business information.

4. Erect a large, legible road sign next to your road gate. If you live far away from "civilization," erect directional signs along the way that lead to your farm. Before mounting them, however, review signage laws and ask permission from landowners.

5. If you live on a well-traveled road, display a small herd of attractive, eye-catching goats in a pasture adjacent to the road or erect a kid playground close to the highway. Nothing stops traffic like a pack of gamboling baby goats.

6. Take out your goats in public. Train a harness or packgoat, then drive or lead it in parades with farm signs attached to its cart or panniers. Take a well-behaved goat on visits to schools and nursing homes. Take goats and a display booth to community gatherings, county fairs, and farm expos.

7. Give goat talks and demonstrations (haul along a friendly goat whenever you can). Many civic groups need speakers for meetings and events; let it be known that you're available and interested. Other speaking possibilities: Hold an open house and sponsor a 4-H meat goat clinic at your farm.

Get out your goats where the public can see them.
Everyone — even a goat — loves a parade.

8. Serve chevon (BBQ sandwiches are always a hit) whenever and wherever you can, especially at public functions and to people who may not have tasted goat meat before. Distribute recipe sheets with your contact information printed on them.

9. Join and participate at goat-related listservs and e-mail lists. It costs nothing but your time and can generate more sales than any other sort of free promotion. People tend to buy from breeders they know, and you can advertise for free on most of these lists.

10. Configure your e-mail program to add a business-related signature to your outgoing e-mail. At a minimum include your farm name, location, a tag line describing your services, and your farm's web address. For example:

Ozark Thunder Boer Goats

Mammoth Spring, Arkansas

"Ennobled bloodlines at reasonable prices"

www.OzarkThunder.com

11. Write a goat column for your local newspaper in exchange for printing your farm name and web address in each article's byline.

12. Hold a contest. Donate a wether to the youngster who writes the best essay on the topic "I would like to show a meat goat in 4-H because . . ." or a plush toy goat to the kindergartners who dream up the cutest names for the twin bottle kids you bring to their school for a visit. (In this instance, be sure to distribute consolation prizes to the rest of the class — perhaps goat-shaped cookies.)

13. Write a winning press release. When you hold a contest or host Goat Day at your farm, when your doe kids quadruplets or your buck becomes ennobled, send or hand-deliver press releases to newspapers in your locale.

Appendix A

DEFRA's Code of Recommendation for the Welfare of Goats

DEFRA IS BRITAIN'S DEPARTMENT OF ENVIRONMENT, Food, and Rural Affairs. Among its duties, DEFRA issues codes for the humane treatment of all livestock species, including goats. This code is the law in Great Britain; to cause unnecessary pain or unnecessary distress to any farm animal is an offense under the Agriculture Miscellaneous Provisions Act of 1968. The department's guidelines are a good guide to goat keeping in other countries too. If you follow the code, you'll have happy, healthy, productive goats — what more could any producer ask for? The following is the code as written by DEFRA.

Code of Recommendation for the Welfare of Goats

Preface

The Code of recommendations for the welfare of goats, which is made under Section 3 (1) of the Agriculture (Miscellaneous Provisions) Act 1968 and approved by Parliament, is intended to encourage all those responsible for looking after these animals to adopt the highest standards of husbandry. It takes account of five basic animal needs: freedom from thirst, hunger and malnutrition; appropriate comfort and shelter; the prevention, or rapid diagnosis and treatment of, injury, disease or infestation; freedom from fear; and freedom to display most normal patterns of behaviour.

The Code is backed up by the law of the land. To cause unnecessary pain or unnecessary distress to any farm animal is an offence under The Agriculture (Miscellaneous Provisions) Act 1968 — the breach of a Code provision, whilst not an offence in itself, can nevertheless be used in evidence as tending to establish the guilt of anyone accused of causing suffering under the Act (Section 3).

Without good stockmanship, animal welfare can never be adequately protected. The Code is designed to help stockmen particularly the young and inexperienced to reach the required standard.[1]

For the purposes of this Code the word *goat* refers to all caprine stock, and an animal for which six months is considered to be a kid.

Introduction

1. Goats in Great Britain cover a variety of breed types, each with its own unique characteristics. The recommendations in this Code are appropriate to goats under various husbandry systems, and their application will help to ensure that the welfare of the stock is safeguarded.
2. The goat has a natural tendency to browse and range for its food and these factors should be taken into account in deciding on a suitable environment. Many breeds of goat require more protection from inclement weather than cattle or sheep and, whatever husbandry system is adopted, some form of shelter should be provided.
3. Goats, being gregarious animals, prefer to live in social groups and appear to enjoy human contact. If kept singly, they require more frequent contact with, and supervision by, the stockman. They should always be treated as individuals, even when kept in large herds. When forming new groups, care should be taken to avoid fighting and stress if adult animals are

mixed (see paragraph 30). Goats prefer to be led but can be driven if care is taken.

4. The number and type of goats kept and the stocking rate should depend on the suitability of the environment and the skills of the stockman.

5. Although very large herds can be managed successfully, in general, the larger the size of the unit the greater the degree of skill and conscientiousness needed to safeguard welfare. The size of a unit should not be increased nor should a large unit be set up unless it is reasonably certain that the stockman in charge will be able to safeguard the welfare of the individual animal.

Health

6. The stockman should know the normal behaviour of goats and recognise the signs which indicate good health. These include good appetite, alertness, good coat condition, absence of lameness, firm round droppings (similar to those of a sheep or rabbit) and no visible wounds, abscesses or injuries. Purchased stock should be healthy and free from infectious disease.

7. Goats should be inspected regularly, particularly for foot condition (see paragraph 50) and parasitic infections of the skin (e.g., lice and mange), to which they are susceptible.

8. The health of the goat should be safeguarded by the appropriate use of preventive measures such as parasitic control and vaccination programmes based on veterinary advice (see paragraph 14).

9. When goats are ill, they soon lose the will to live. The stockman should identify the cause of the goat's deterioration, should separate injured or ailing goats and take immediate remedial action. Prompt veterinary advice should be obtained if the goat appears to be seriously ill or in pain, the cause of the deterioration is not clear, or if the stockman's action is not effective.

10. If a goat has to be destroyed on the farm, this must be done humanely and, where possible, by a person who is experienced in both the technique and the equipment used for slaughtering goats.

Feed and Water

11. Goats should receive daily a balanced diet which is adequate to maintain full health and vigour. They should have access to sufficient fresh, clean, water at all times. If this is impossible for any reason, such water should

be provided at least twice daily. Goats prefer water which is not excessively cold.

12. Feed should be palatable and should be placed in suitable racks or containers. Stale and fouled food should be removed.

13. Goats need a comparatively large quantity of bulky feed. They have a preference for coarse forages and tree branches. Suitable foods for housed goats include pea and bean haulm [the stems of peas, beans, potatoes, and grasses], clover, lucerne and meadow hay and silage(s) and coarse, flaky or pelleted concentrated food. Care should be taken not to overfeed certain foods, for example concentrates, as this can lead to such problems as bloat, acidosis, laminitis and obesity.

Grazing

14. Grazing should include a variety of plants to ensure an adequate intake of roughage and minerals. If grazing is poor, supplementary feeding may be required. Goats should be moved at appropriate intervals to clean pastures to control parasite infestation, and this should be combined with a regular parasite control programme (see paragraph 8).

15. Being browsing animals, goats should be denied access to poisonous shrubs, trees and plants within grazing areas. Well-known examples that are poisonous are rhododendron, yew, laurel and bracken, but there are many others and expert advice should be sought where necessary.

Fencing

16. Goats have a tendency to jump and clamber. Fencing should be strong enough and of sufficient height (at least 1.2 metres) to prevent them from escaping. It should be designed, constructed and maintained so as to avoid the risk of injury.

17. Electric fences should be so designed, installed and maintained that contact with them does not cause more than momentary discomfort to the goat. Electric mesh type fences are not suitable for horned goats and young kids.

Tethering

18. Outdoor tethering, if carried out, requires a high degree of supervision, with inspections at frequent intervals. Tethered goats are particularly vulnerable to worrying by dogs and teasing by children. Goats should not be tethered where there are obstacles and a risk of the chain becoming

entangled. Tethers should be designed and maintained so as not to cause distress or injury to the goats. Collars should be light but substantial and attached to a strong chain not less than 3 metres in length with at least two swivels. Particular care should be taken to provide food, water and shelter.

19. Kids should never be tethered.

Housing, Buildings and Equipment

20. Advice on welfare aspects should be sought when constructing and modifying buildings. The lying area should be covered, dry and well-lit with sufficient ventilation which does not cause draughts at animal level. Goats are very inquisitive and all gate / door fastenings should be goat-proof.

21. Fittings and internal surfaces of all buildings and equipment to which goats have access should not have sharp edges or projections. Fittings should be so arranged as to avoid injury.

22. Surfaces should not be treated with paints or wood preservatives which may cause illness or death. There is a risk of lead poisoning from old paint work, especially when second-hand materials are used.

23. Hay racks and nets should be properly positioned and designed to avoid the risk of injury, in particular to the eyes of all types of goats. Hay nets should not be used for young kids and horned goats as there is the danger of them becoming entangled.

24. When goats are fed in groups, there should be sufficient trough space or feeding points to avoid undue competition for food.

25. Water bowls and troughs should be constructed and sited so as to avoid fouling and to minimise the risk of water freezing in cold weather. They should be kept thoroughly clean and be checked at least once daily, and more frequently in extreme weather conditions, to ensure that they are in working order.

26. Floors should be designed, constructed and maintained so as to avoid discomfort, distress or injury to the goats. Solid floors should be well drained. Sufficient clean, dry bedding incorporating straw or other suitable material should be provided to ensure comfort and reduce the risk of injury to the udder.

27. If housed, male goats should be within sight and sound of goats or other animals and in strongly constructed buildings which allow sufficient room for exercise.

28. Housed goats should have access to a yard or pasture.

29. The space allowance when penned should be calculated in relation to the age, size and class of stock. This and the size of the group should be based on appropriate advice. Horned and polled goats should not be put in the same pen unless reared together.

30. The introduction of a new goat or goats to an existing group can result in bullying. This may be alleviated by increasing the space allowance or by penning the new animal adjacent to the existing group for a short period.

31. All electrical installations at mains voltage should be inaccessible to goats, well insulated, safeguarded from rodents, and properly earthed.[2]

Mechanical Services and Equipment

32. All equipment and services, including drinkers, milking machines, ventilating fans, heating and lighting units, fire extinguishers and alarm systems, should be kept clean, inspected regularly and kept in good working order.[3] Alternative ways of milking and maintaining a satisfactory environment should be available in case of failure.

Pregnancy and Kidding

33. Heavily pregnant females should be handled with care to avoid distress and injury which may result in premature kidding.

34. Pregnant and nursing females should receive sufficient food to maintain the health and bodily condition of the goat and ensure the development of healthy kids. This is particularly important during the last 6 weeks of pregnancy. Water should always be available.

35. Stockmen should pay particular attention to cleanliness and hygiene. Every effort should be made to prevent the buildup and spread of infection by ensuring that kidding pens are provided with adequate clean bedding and are regularly cleansed and disinfected. A kidding pen within sight and sound of other goats is desirable. Any dead kids should be removed without delay.

36. Stockmen should be sufficiently familiar with problems arising at kidding to know when to summon help. Veterinary advice should be sought when the need arises.

37. It is vital that every newly born kid receives colostrum from its dam or from another source as soon as possible and in any case within 6 hours of birth. Adequate supplies of colostrum should be stored for emergencies, but pooled colostrum, for example, from other premises, may constitute a disease risk.

Artificial Rearing

38. Artificial rearing can give rise to problems and, to be successful, requires close attention to detail and high standards of supervision and stockmanship. Particular attention should be paid to cleanliness and hygiene.

39. Young kids should always have access to milk substitutes or be fed at least 2 or 3 times each day. Milk from other dams could constitute a disease risk. Fresh fibrous food should be available from 1 to 2 weeks of age.

40. Some form of safe supplementary heating, particularly in the early days of life, may be necessary.

41. A dry bedded lying area and adequate ventilation should be provided at all times.

Disposal of Unwanted Kids

42. Unwanted kids should be treated as humanely as those being kept for rearing, and if they are to be killed, arrangements should be made for this to be done as humanely as possible (see paragraph 10).

Milking

43. The stockman should be aware of the specific problems of a lactating goat and the ways in which these problems can be avoided or alleviated. Veterinary advice should be sought where necessary.

44. Special attention should be paid to milking techniques so that injury to teats can be avoided. Good milking practices should include careful handling, an examination of foremilk and the avoidance of excessive stripping.

45. Before and after milking, hygiene measures should be adopted to reduce the spread of disease. Failure to attend to hygiene and to the efficient functioning of milking machines can lead to mastitis and damage to teats.

46. Goats can milk through to 24 months, but this should be supported by adequate nutrition (see paragraph 11).

47. Lactating goats should be milked daily or sufficiently often according to yield.

Milking Parlours and Equipment

48. Pens, ramps, milking parlours and milking equipment should be properly designed, constructed and maintained to prevent injury and distress.

49. It is essential to ensure that milking machines are functioning correctly by proper maintenance and adjustment of vacuum levels, pulsation rates and ratios, taking account of manufacturers' recommendations.

Foot Care

50. Close attention should be given to the condition of the feet, and, where necessary, regular foot trimming should be carried out. Goats should be kept in accommodation which is dry underfoot.

Disbudding and Dehorning

51. These operations must be carried out by a veterinary surgeon.[4] If disbudding is to be carried out, this should be done at the earliest possible age; 2–3 days is ideal, but not later than 10 days. Dehorning an adult goat is a stressful procedure and should be avoided.

Castration

52. Castration, if necessary, must be carried out by a trained operator in strict accordance with the law.[5]

Shearing and Combing (Fibre Production)

53. When shearing, care should be taken not to nick or cut the skin. Where a wound does occur, immediate treatment should be given.
54. The goat is particularly susceptible to changes in temperature. Unless housed, goats should be shorn only in suitable weather conditions. Combing is preferable to shearing in adverse weather conditions.
55. Protection by housing or by the use of a coat should be provided if inclement weather occurs after shearing.

Fire and Other Emergency Precautions

56. Stockmen should make advance plans for dealing with emergencies such as fire, flood or disruption of supplies, and should ensure that all staff are familiar with the appropriate emergency action. At least one responsible member of staff should always be available to take the necessary action.
57. Fire precautions should be a major priority for the good stockman. The provisions of Section 1.3 of British Standard BS5502 should therefore be followed. Expert advice on all fire precautions is obtainable from fire prevention officers of local fire brigades and from the Fire Prevention Association.
58. In the design of new buildings or alteration of existing ones there should be provision for livestock to be released and evacuated quickly in case of emergency. Materials used in construction should have sufficient fire

resistance. Adequate doors and other escape routes should be provided to enable emergency procedures to be followed in the event of a fire.

59. All electrical, gas and oil services should be planned and fitted so that if there is overheating, or flame is generated, the risk of flame spreading to equipment, bedding or the fabric of the building is minimal. It is advisable to site power supply controls outside buildings. Consideration should be given to installing fire alarm systems which can be heard and acted upon at any time of the day or night.

60. In case a 999 call has to be made, notices should be prominently displayed in all livestock buildings stating where the nearest telephone is located. Each telephone should have fixed by it a notice giving instructions for the Fire Brigade on how to reach the buildings where the goats are housed.

References

1. Training courses which follow Code recommendations are arranged for stockmen by the Agricultural Training Board, Agricultural Colleges and local educational authorities. Proficiency testing in relevant subjects is carried out in England and Wales by the National Proficiency Tests Council, and in Scotland by the Scottish Association of Young Farmers' Clubs.

2. Any installation or extension involving mains electricity should comply with the Regulations for the Electrical Equipment of Buildings issued by the Institute of Electrical Engineers.

3. The Welfare of Livestock (Intensive Units) Regulations 1978 (SI 1978 No 1800) requires stockkeepers of intensive units to inspect their livestock and the equipment upon which such stock depend at least once daily.

4. Under the Protection of Animals Acts 1911 to 1988 (in Scotland, the Protection of Animals [Scotland] Act 1912 to 1988), it is an offence to castrate a goat, which has reached the age of 2 months without the use of an anaesthetic. Furthermore, the use of a rubber ring or other device to restrict the flow of blood to the scrotum is permitted without an anaesthetic only if the device is applied during the first weeks of life. Under the Veterinary Surgeons Act 1966, as amended, only a veterinary surgeon or veterinary practitioner may castrate a goat after it has reached the age of 2 months, or dehorn or disbud a goat, except the trimming of the insensitive tip of an ingrowing horn which, if left untreated, could cause pain or distress.

5. See Note 4, above.

Appendix B

Photograph
Your Goats

WHETHER IT'S FOR YOUR WEBSITE, your business card, or a magazine ad, a bad photo of your goats is worse than no photo at all. In many cases, potential buyers' initial contact with your goats will be through photos. Because there are few goat-savvy professional photographers, it pays to learn

FREE PUBLICITY: PICTURE THIS

Book and magazine publishers who are looking for photos or illustrations often contact breed registries for specific images or general research material. Ask your breed registry what kind of images they accept. Most organizations prefer digital images these days, but may have requirements as to size and dpi (at least 350 is typical). Others will accept prints but may want them to be at least 5 × 7. For prints, attach the necessary identification on a piece of paper taped to the back (ink often bleeds through). For digital images, name each file appropriately (for example, Downen R33 "Hoss" enobled, owned by MAC Goats, Winona, MS). You'll be surprised where your photos turn up — maybe in a book such as this one!

to take great pictures yourself. And it isn't hard; here's what you need to know to get started.

What Kind of Camera? What Kind of Film?

If you're taking pictures for your website, you'll need digital images. You can shoot them using a digital camera or scan print images into digital format. Because you want your pages to load quickly, you'll need low-resolution (lo-res) images (75 dpi, or dots per inch, is a good size for websites), and today's simplest digital cameras handle low-resolution pictures with ease.

Higher resolution (hi-res) digital pictures work well for use on business cards, flyers, and the like, but if you need pictures for a magazine ad (or article), check with the publication before submitting digital images. You'll need a high-end digital camera to shoot the sharp, high-resolution images (at least 300 dpi in large formats) needed for publication. If you have a suitable camera, by all means use it. If not, supplying 35 mm images is a better choice. If you do shoot 35 mm prints or slides, chose 100 or 200 ASA film; it's what most editors demand.

If you have neither a digital nor a 35 mm camera, fear not; you can still shoot decent pictures with an inexpensive camera if you understand its limitations (we'll talk about that in just a moment).

Preparation Is the Key to Great Posed Portraits

Before shooting portraits, teach your goat to pose. There are few things more frustrating for a photographer than trying to snap great images of a goat that refuses to stand still.

First, spiff up your goats before shooting their portraits. Pretend they're going to a show. An image of a clipped, bathed goat standing in a posed stance makes a terrific first impression.

Here are a few other tips:

- Strive for simplicity. Breathtaking scenery is fine in its place, but you want to draw attention to your goat.
- Avoid busy backgrounds (distant buildings, vehicles, fence posts or utility poles, a mass of other goats milling around) and clutter (board piles, disintegrating fences, buckets or other junk strewn on the ground).
- Pose a light-colored goat against a dark background (a solid bank of trees is nice) and a dark one against light (sky, for instance).

This Boer buckling, whose hindquarters are growing faster than his shoulders (typical in goatlings of his age), pauses slightly uphill, thus helping to mask his temporary "downhill" conformation.

- Choose a level spot or one where the goat can stand with its forequarters very slightly uphill.
- Recruit help. Unless you photograph your goat at liberty (loose in a pen or pasture with no handlers involved), you'll need a goat-savvy assistant to pose it (ask him or her to dress as though going to a show — he or she may be in the photo). You'll need a second assistant to grab and hold the goat's attention, especially if you're shooting a Boer and you want to highlight the goat's handsome profile.

On portrait day, pack along extra memory cards for your digital camera or bring plenty of film and expect to shoot plenty of frames. Even professional photographers are happy when they shoot one really great image per session. And be sure to allow plenty of time to capture that perfect shot — one or two hours, at least.

Posing a Buck

One of the best (and easiest) ways to shoot a "posed" picture of a buck is to turn him loose in an area with great backgrounds and wait until he has to pee. He'll stand stock-still with his feet squared and his head held high long enough for you to get several shots. Examine good pictures of bucks with this in mind and you'll see how often it's done.

Attention-Grabbers Save the Shoot

No two goats respond to the same attention-getting ploys and props, but these consistently work for us. Collect a bag of props to keep with your camera and use them to shoot great pictures.

- **A kazoo.** Tootling on a toy kazoo is guaranteed to capture just about any animal's attention. Start slowly; the odd vibrato sound of a kazoo initially frightens some goats.
- **A plastic bag on a stick.** Wave it around. It works!
- **An umbrella.** To avoid strong reactions, open it *slowly* the first few times.
- **Additional noisemakers.** A dog's squeaky toy works for flighty goats that spook at umbrellas, kazoos, and plastic bags. A spring-loaded steel measuring tape, a child's toy that makes a buzzing sound, a baby rattle — any sort of unusual noisemaker tends to fascinate most goats.
- **A bucket with a few handfuls of grain in the bottom.** Shake the bucket, show the feed to the goat (or give the animal a taste), then shake that bucket some more.
- **Things to throw.** When you want the goat to gaze at one specific spot, toss an object at the spot to capture his attention. A noisemaker (such as a jingle ball designed for dogs or a dozen dimes in a resealed soda can) works best.

If you don't have props, improvise. Ask your helper to jump, hoot, toss his hat in the air, skip in a circle, or roll on the ground. Goats aren't accustomed to seeing us behave this way, so they tend to stand and stare in wonder when humans do such things.

Or bring around another goat. A rival buck will cause the one you're photographing to strike a magnificent pose.

Three-quarter rear views showcase a meat goat's hindquarters — but not when his head is lowered to graze. Having an assistant to grab this buckling's attention would have resulted in a much better image.

If you'd prefer to emphasize a buck's powerful hindquarters, move slightly (about 45 degrees) toward his rear before snapping the shutter. Unless you have a telephoto lens and know how to use it, however, it's best to avoid shooting three-quarter or greater hind views; while they do emphasize massive hindquarters, taken with a standard lens (or especially with a single-use camera) they'll distort the goat's proportions and make his front end seem tiny.

Let There Be Light

Regarding light when you photograph, here are some tips to remember:

- **Shoot with the sun at your back or slightly over one shoulder.** Never shoot facing directly into the sun.
- **If your camera has an automatic flash, use it.** Flash fills in shadows, even on a fairly sunny day.
- **Take photos on a slightly overcast day.** Lighting will be softer, color saturation will be vastly improved, and you can shoot from any angle because you won't have to worry about keeping the sun at your back.
- **Never shoot pictures when the sun is directly overhead.** Harsh, dark shadows obscure the subject's sides and belly. Early to mid-morning and mid-afternoon to early evening work best for taking photos. Many professionals favor 10 A.M. and 2 P.M. for grabbing high-muscle-definition shots.

Position Yourself to Shoot Stellar Photos

The trick to shooting great goat photos is to kneel or squat to the animal's level. *Don't* stand and shoot down at the goat; it distorts the animal's proportions by shortening its legs and making it appear dumpy — not at all a flattering pose!

To shoot great side views, position yourself so the goat fills the frame and your camera is level with and pointing at the center of the animal's barrel. You want to see all four legs. The cannon bones in the legs closest to you should be perpendicular to the ground; the legs on the opposite side should be closer together, so you can see them under his belly.

Here's a great "personality" shot as a buck peeks through
the bars of a gate.

Sometimes Great Pictures Simply Happen

Outstanding candid photos can be wonderful promotional tools too. A newborn goat kid snuggled against its mama's side, two brawny bucks squaring
off for a fight, a passel of Boer babies racing with their long ears flying in the
wind — editors love these kinds of great candid pictures. So take along your
camera as you stroll among your goats — and be prepared; photo ops happen
all the time.

Appendix C

Identify Your Goats

GOATS MUST BE TATTOOED BEFORE most registries will accept applications for their registration, certain classes of goats must be identified with official ear tags to comply with the USDA's mandatory scrapie eradication program, and if the National Animal Identification System is federally approved (it's pending as this book goes to press), all goats will be ear-tagged or microchipped as a matter of course. Identification is an increasingly important part of raising goats.

Tattoos

Applying tattoos sounds scary but the process is easier than you might think. First, assemble the correct tools.

- **Tattoo pliers with alphabetical and numeric digits.** Before you join most goat registries, you must apply for an official herd prefix of two to five letters/digits. You'll have to tattoo that series of letters and/or numbers in one ear (with most registries, the right one) along with a letter denoting the goat's year of birth (check with your registry to be certain you know which it is) as well as an individual identification number of your choosing in the opposite ear. Tattoo pliers come in ³⁄₁₀-inch-, ⁵⁄₁₆-inch-, and ³⁄₈-inch-digit sizes. If you have a five-digit herd prefix, make sure the set you choose accepts five digits (some don't). Better pliers have a positive ear release feature; pay a little extra and get it — you and your goats will be glad you did.

- **Tattoo ink.** Tattoo kits come with a jar of black ink. Virtually every goat producer we've talked to agrees: Throw it away and buy green paste ink. Green is better than black because it shows up well in dark ears and paste is best because the roll-on ink tends to drip.
- **An old, soft toothbrush.** You'll use this to scrub ink into the new tattoo.
- **Paper to test the digit setups before tattooing the goat's ear** (index cards work very well). If you keep hard-copy records, "tattoo" each goat's individual record to indicate that the animal has been tattooed. Some producers stamp each animal's registration certificate as well, but check with your registry before doing it; with some registries, defacing registration papers renders them null and void.
- **Alcohol and paper towels or cotton balls.** Thoroughly clean ears prior to tattooing them.
- **Disposable gloves.**
- **A fitting or milking stand.**
- **A helper.** (If you don't have a fitting or milking stand, you'll need two.)

To tattoo a goat, assemble all your tools before you begin. Sterilize the tattoo digits by immersing them in alcohol or a solution of one part household bleach to five parts water, then place them in the pliers. If you use the bleach solution, follow it with a plain-water rinse just prior to tattooing the goat. Then follow these steps:

1. Test the tattoo on paper. Make certain you've inserted the series correctly and the digits are right-side up and facing the right direction.
2. Put the goat on your fitting or milking stand if you have one; otherwise, halter or collar and tie it where you can easily access its ears. A little feed will help distract most goats, at least until you clamp the pliers on their ear.

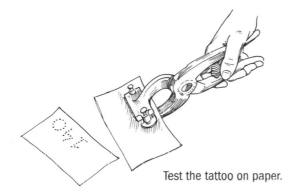
Test the tattoo on paper.

3. Use paper towels or cotton balls and alcohol to *thoroughly* swab the inside of the goat's ear. Pat it dry.

4. Now have your helper hold the goat's head in place. He or she should stand out of your way while supporting the animal's chin with one hand and grasping a horn or the animal's poll with the other. If you aren't using a stand, have a second helper crowd the goat against the wall to help keep it in place.

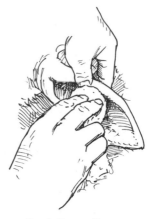

Swab the goat's ear.

5. Flatten the ear as best you can. Position the pliers in the ear, making certain the digits are inside the ear and you aren't holding the pliers upside down. Avoid tattooing into blood vessels, ridges of cartilage, and scar tissue. Now press quickly but *firmly* and then release — this is your only chance to do a good job, so do it right. If the needles stick to the ear (or even pierce through to the other side), peel the ear away and apply less pressure the next time.

6. If some of the punctures bleed (and they probably will), pinch a paper towel or cotton ball over the bleeders until they stop. Don't apply ink until the bleeding has ceased; blood flow washes ink out of the piercing holes.

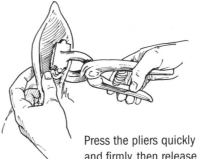

Press the pliers quickly and firmly, then release.

7. Next, squirt a blob of ink onto your finger, making certain you get pigment, not just oil, out of the tube. Rub ink into the piercings with your finger or (better yet) a soft-bristled toothbrush, taking care to get ink into each and every one. When they're filled, leave the excess ink on the ear (scrubbing it off may dilute ink in the piercings as well) and proceed to the other ear.

8. When you're finished, it never hurts to give the goat a shot of tetanus antitoxin. Then remove and disinfect the digits before moving on to another goat. At day's end, disinfect everything — pliers, toothbrush, and all — before storing these supplies.

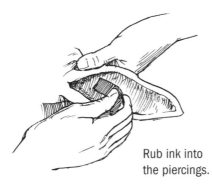

Rub ink into
the piercings.

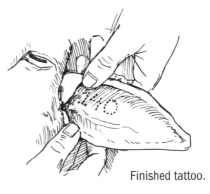

Finished tattoo.

Tattooing Tips

- "Right ear" and "left ear" refer to the goat's right and left. Stand at the animal's rear, facing forward, if you get confused.

- Tattoo at least a month before a show to allow tattoos to heal and be fully legible.

- Tattoo needles must make holes that are large enough to accept an adequate amount of ink or the tattoo will be difficult to read. Brand-new, ultra-sharp digits don't accomplish this. *Lightly* filing the tips before using new digits often helps.

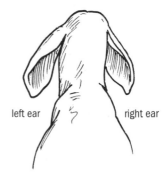

left ear   right ear

- Use a small-digit pliers set when tattooing kids; the tattoo will get bigger as the ear grows.

- When tattooing a LaBoer (a LaMancha and Boer cross), you'll tattoo its tail web instead of its ears (LaManchas' tiny ears aren't conducive to tattooing), but the process is the same. Place the left-ear tattoo on the left side of the tail web and vice versa. You'll have to squeeze a little harder to make a good impression.

- If a tattoo fades, you'll have to retattoo that ear. Check with your registry regarding its policy on reapplication of tattoos.

- When it's difficult to read the tattoo in a dark-pigmented ear, scrub the ear with alcohol to remove grease and dirt and then shine a flashlight behind the outside of the ear. You'll be surprised how well this works!

Ear Tags

The USDA's mandatory scrapie eradication program is overseen on the state level, so depending on where you reside, you may not have much choice in the type of ear tags you're required to use. When you can, however, choose your ear tags wisely: double-sided, button-type plastic tags and metal tags tend to cause infections, so avoid using them. Choose small, two-piece plastic models that swivel — they're far less likely to snag on fencing or brush and tear the ear.

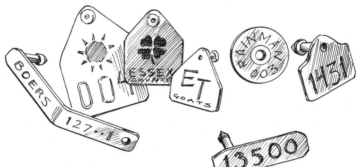

Ear tags come in a variety of types and shapes.

To properly install an ear tag, follow these steps.

1. Restrain the goat using the same protocol used for tattooing.
2. Clean the ear thoroughly, using plenty of alcohol, and then pat it dry.
3. Apply antiseptic salve to the male tip of the ear tag just prior to insertion.
4. When placing the tag, avoid blood vessels, ridges of cartilage, and scar tissue. The dots on the diagram indicate the best spots (one-piece loop tags fit only when installed in the lower spot). Put the female half of the tag on the inside of the ear.
5. Press quickly and firmly, and it's in!

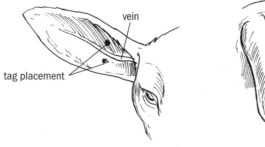

vein

tag placement

The ear tag is in.

Ear Tag Tips

- Use a different color for each birth year or tag the offspring of each buck with a different color to determine a goat's age at a glance.
- Some colors are easier to read than others. When visibility is an issue, in descending order of readability choose yellow, white, orange, light green, black, pink, purple, gray, brown, red, medium or dark green, then blue.
- When using write-on tags, use a marking pen designed specifically for that purpose. Mark the tags the night before tagging to give them time to dry overnight.

Temporary On-Farm ID

Colored plastic neck chains can be used to identify classes of goats according to age, ancestry, kidding dates, productivity and so on — and you can use them to lead tame goats. To reduce the risk of collared goats hanging themselves on fencing, browse, or another goat's horns, choose lightweight plastic chain that breaks easily in an emergency. Secure it, however, with a chain link designed for use with larger, Boer-size plastic neck chains (small-size chain links come apart very easily and your goats will shed their chains faster than you can replace them).

For very temporary markings, use chalk or wax markers, paint brands (inexpensive numerical branding irons dipped in marking fluid), paint sticks, or aerosol-spray paints designed for use on livestock.

Chalk or paint stick markings last about a week, wax markings last about four weeks, and aerosol-spray paint markings or those from paint brands last up to six weeks. All such markings are easy to see from a distance and are great for identifying animals in a herd as you vaccinate or worm them. Some producers use spray paint or paint brands to mark does and kids with matching numbers before these pairs leave the mothering pen. The first family is marked with a 1, the second with a 2, and so on. This has the added advantage of giving the producer the approximate age of each kid at a glance.

Chalk, paint stick markings, wax, and aerosol paint markings are all temporary but easy to see for short-term identification.

Appendix D

Trim Your Goats' Hooves

HOW OFTEN YOU'LL NEED TO TRIM your goats' hooves depends on their breed and where and how you keep them. The hooves of dairy goats kept in drylots or Boers raised on boggy meadows in the North will require more frequent trimming than those of Kikos browsing rocky, rugged Ozark hills. The former may require trimming every other month; the latter, possibly never.

The coronary band (the area where hoof meets leg) of a properly trimmed goat is almost parallel to the ground (look at a newborn kid's hooves to see what hooves should look like). Front hooves wear down faster, so to know when to trim a goat's hooves, evaluate the back hooves first. If a hoof is angled out toward the front, it needs trimming.

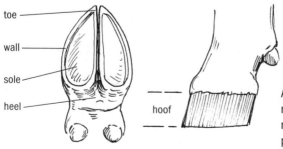

A newly trimmed hoof should resemble the drawing on the near left. The hair line should be parallel to the ground (right).

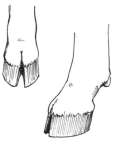

Before trimming   After trimming

The Tools of the Trade

Hoof trimming is easy, given the right equipment. There are several combinations of hoof-trimming tools that do the trick.

■ **Goat and sheep trimming shears.** These are the first choice of most goat owners. The old standby is the orange-handled Shear Master trimmer, and its relatively new competitor is Burgon and Ball's green-handled EzeTrim hoof shear. Both are very sharp and reasonably priced. Because these tools resemble garden pruners, some goat owners substitute pruners for the real McCoy, but pruners aren't as sharp and they're harder to handle. For the price, real hoof shears are much better tools.

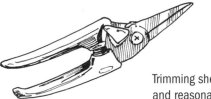

Trimming shears are very sharp
and reasonably priced.

■ **Foot rot shears.** These are heftier versions of the orange- or green-handled trimming shears. Some have serrated blades and some have special rotating handles that are said to reduce hand fatigue.
■ **Power shears.** Large-scale producers who prefer ease of operation paired with speed choose these. The drawback is their cost: several hunderd dollars, and that doesn't include the air compressor needed to run them.
■ **Short-handled horse hoof nippers, a hoof knife, and a rasp.** Producers who trim horses' hooves often prefer this combination over any sort of shears.
■ **Gloves.** You'll need them — trust me.

- **A stiff brush.** Use to clean off the bottom of each hoof.
- **Blood-stop powder.** This is important if you snip a hoof too short (it happens to everyone, so be prepared). Tip: Some producers prefer cornstarch or plain old flour, and even a big wad of cobwebs will stanch the flow of blood if there's nothing else at hand.

Hoof-cleaning brush

How to Proceed

Trimming doesn't hurt the goat unless you "quick" the hoof by cutting into living tissue, and if you're careful, that very rarely happens. Assuming you'll use the standard orange- or green-handled trimmer, here's what to do.

1. Restrain the goat. If you have a fitting or milking stand, use it. If not, halter or collar the goat and secure it next to a wall.
2. If you're using a stand, begin with whichever hoof you choose; otherwise, start with the front leg farther from the wall (that way, if the goat fusses, you can lean into the animal and the wall will support it).
3. Hold the leg so it flexes back at the knee or hock, hoof facing up, with your hand supporting the goat's pastern. Don't twist the goat's leg out to the side; this hurts, causing the animal to fuss.
4. If the hoof is dirty, scrub it with a stiff brush.

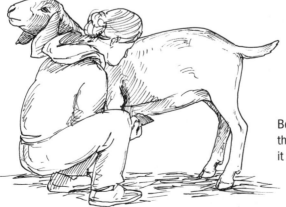

Begin with the front leg farther from the wall, holding it so it flexes at the knee.

5. Using the point of your trimmers or a hoof pick, dig out any dirt stuck in the hoof.

6. Snip off any folded-over or excess hoof wall until it's even with the sole all the way around.

7. Scoop out or trim away any rotted pockets between the sole and the hoof wall.

8. Trim the toe so it's even with the sole of the hoof.

9. Snip off any excess growth between the two heel areas.

10. Trim the sole if it needs it; go very slowly, a sliver at a time, and stop when the sole shows a hint of pink.

11. Some people finish with a sweep or two of the sole using a hoof rasp (this makes the job look neater).

12. Finally, check the dewclaws. If they're excessively long or ragged, carefully snip them back a sliver at a time until they're nice and neat. This is rarely necessary.

Wall

Dig out dirt

Trim wall

Trim heels

Trim nail walls, removing any overgrown or ragged hoof wall.

Trim excess hoof growth here.

dew claws

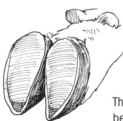

The finished hooves should be perfectly level.

Hoof-Trimming Tips

- Dry hooves, especially hooves that are extremely dry as a result of an ongoing drought, are so hard that they're exceedingly difficult to trim. For best results, trim them when hooves are moist from morning dew, or build a hoof-soaking spot in a small pen and confine goats to the area for a few hours before you trim their hooves.

- When trimming badly overgrown hooves, do it in increments. Trim them as best you can, then wait two weeks and trim some more, continuing the process until the hooves are in tip-top shape. This reduces the chance that you'll make them bleed and results in a neater-looking job.

- Improve minor flaws in the leg conformation of show goats through judicious corrective trimming. Trim the toes shorter (to an angle greater than 45 degrees) on a sickle-hocked hind leg; for a post-legged (too straight) hind leg, leave more toe. If a goat toes out in front, trim the entire inner claw slightly shorter and lower than the outside half. For splayed claws, trim the inner walls of each claw shorter than the outer side. But *always* make changes over the course of several trimmings: abrupt changes in angles can hurt a goat and make it lame.

- If you slice into a blood vessel and the hoof keeps spurting blood, cauterize the blood vessel with a hot iron or pinch it with the hemostat from your first-aid kit.

- Should you encounter hoof rot (you'll know it by its crumbly texture and horrendous smell), trim out all the rotted tissue, even if you have to pare away some hoof wall. Then pour peroxide in the cleaned area and consult your vet or goat mentor for directions on how to properly treat this infectious disease.

Appendix E

A Milk Goat for the Kids

UNLESS YOU SELL OR GIVE AWAY YOUR ORPHANS and rejected kids, you're going to have bottle babies to raise. While today's milk replacers do an admirable job of replicating doe's milk, nothing works better than the real McCoy, so many meat goat producers keep a dairy doe (or two) to provide real goat milk for their annual crop of bottle kids.

And it's not just about making milk. When bred to meat breed bucks, dairy does produce excellent market kids. Well-built 50 percent and higher Boer doelings out of high-quality Nubian does sell for a pretty penny as percentage show goats. A dairy doe pays her way twofold. You might want to buy one for your farm.

Breeds

There are seven full-size dairy breeds available in North America: Alpines, LaManchas, Nubians, Oberhaslis, Saanens, Sables, and Toggenburgs. There are also miniature versions of all seven breeds, mid-size Kinder goats (a breed created by crossing Nubians and Pygmy goats), and two natural miniatures: Nigerian Dwarf and Pygmy goats. If you're raising meat goats too and intend to cross-breed, it's best to stick to one of the full-size dairy breeds. While some folks successfully breed meat goat bucks to mid-size and miniature does, it's not a good idea due to increased incidence of kidding problems (and those cute little girls don't give nearly as much milk as their full-size kin).

THE DAIRY BREEDS

Breed	Origin	Color	Size
Alpine	French Alps (but considered a Swiss breed)	Several recognized colors; every color except solid white or brown with white Toggenburg-like markings	Medium-large; does at least 30 inches tall
LaMancha	United States (California)	Any color or pattern	Medium-large; does at least 28 inches tall and 130 pounds
Nubian	England (where the breed is called Anglo-Nubian	Any color or pattern	Medium-large and elegant; does at least 30 inches tall and 135 pounds; meatier than other dairy breeds
Oberhasli (pronounced oh-ber-HAAS-li)	Switzerland	Chamois (bay-colored) with black dorsal stripe, udder, belly, and lower legs; head is nearly black with two white stripes on each side	Medium; does at least 28 inches tall and 120 pounds
Saanen (pronounced SAH-nen)	Switzerland	White or pale cream	Medium-large; big-boned
Sable	Switzerland	Colored version of Saanen; now considered a separate breed	Medium-large; big-boned
Toggenburg	Switzerland	Any shade of brown; white stripes on either side of face; white stripes on inside of lower legs; white stripes on sides of tail	Medium; big-boned

Ears	Coat	Profile	Lactation Range per Year in Pounds	% Butterfat in Milk
Erect	Short to medium	Straight or dished	840–5,300	3.5
Absent or very small external ears; gopher ears: up to 1 inch with no cartilage; elf ears: up to 2 inches but with cartilage	Short and glossy	Straight	1,050–3,510	3.9
Long, pendulous; ideally reach ½ inch beyond muzzle	Short, glossy, and fine	Convex (Roman-nosed)	640–3,670	4.8
Erect	Straight	Straight or slightly dished	990–3,630	3.5
Erect; medium, preferably pointing outward	Short	Straight or slightly dished	970–5,630	3.4
Erect; medium, preferably pointing outward	Short	Straight or slightly dished	970–5,630	3.4
Erect; medium, preferably pointing outward	Short to medium	Straight or dished	860–4,480	3.2

ALPINE

SAANEN

LAMANCHA

OBERHASHI

NUBIAN

TOGGENBURG

All of the breeds listed in the chart on pages 288–289 give plenty of milk to feed multiple bottle kids. Excess milk can be frozen or canned for future use or processed for your family's enjoyment.

Milking a Goat

Milking a goat is easy and quick — a proficient milker can hand-milk a mannerly dairy goat in five to seven minutes. It does take practice, however. Invest in a mellow doe that has been previously hand milked. A novice milker and a cranky or frightened doe is a recipe for disaster.

Before milking, assemble the necessary tools:

- **Clean hands with short fingernails**
- **Something to milk into.** An easy-to-sanitize, seamless, stainless-steel milking pail is the tool of choice (and the *only* thing to use if your family will be drinking some of the milk), but if the milk is headed for bottle babies, any widemouth container you can sanitize will do.
- **Udder wash and paper towels** (or unscented baby wipes)
- **Teat dip and a teat dip cup** (or a disposable 3-ounce paper cup) or a spray-on product such as Fight Bac
- **A strip cup**
- **A sturdy milking stand** set up against the wall in your milking area, with a feed of grain waiting in the feed cup

Here's the Step-by-Step for Milking

1. Lead the doe to the milking stand and secure her head in the stanchion.
2. Wash her udder and dry it with a paper towel (or wipe it clean with a baby wipe), then massage her udder for 30 seconds to facilitate milk letdown.
3. Squirt the first few streams of milk from each teat into the strip cup and examine the milk from each teat, looking for watery consistency and strings or lumps (these are early signs of mastitis).
4. Place the milking pail slightly in front of the goat's udder, sit down, and take a teat in each hand.
5. Trap milk in the teat by wrapping your thumb and forefinger around its base. Squeeze with your middle finger, then your ring finger, and then your little finger in one smooth, successive motion to force milk trapped in the teat cistern into your pail. Never *pull* on the teat!

6. Relax your grip to allow the teat cistern to refill and repeat the same squeezing motion. Alternate, squeezing one teat while the other refills.

7. As the teats deflate and become flaccid, gently bump or massage the goat's udder to encourage additional milk letdown. *Don't* finish by stripping the teats between your thumb and first two fingers; it hurts the doe and it isn't necessary.

8. Pour enough teat dip into the teat cup (or paper cup) to dip each teat in fresh solution (don't reuse it!) and allow the teats to air-dry — or spray each teat with Fight Bac or another aerosol teat treatment product.

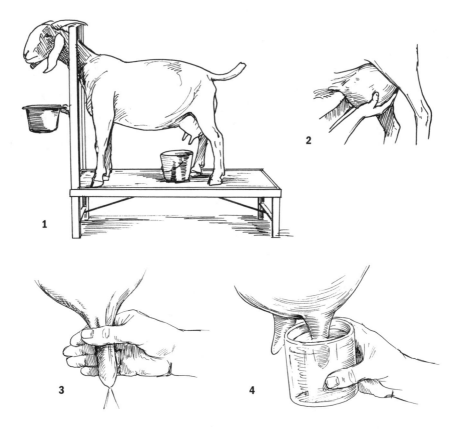

(1) Secure the goat in the milking stand with a pail positioned under the udder. **(2)** After cleaning the udder, massage it. **(3)** Be sure your hand position on the teat is correct. **(4)** After milking, dip each teat in disinfecting teat dip.

Feeding a Dairy Doe

Because dairy does give so much milk, they require considerably more high-quality feed than do comparably sized meat breed does.

At the height of her lactation, a typical doe that's in good flesh should be offered as much clean, palatable hay as she'll eat, a free-choice mineral supplement designed for use with her hay (a high-phosphorus mix for use with legume hay; a two-to-one calcium–phosphorus mix if she's fed grass hay), and one pound of grain or pelleted feed for every three pounds of milk she produces.

Sixteen to 18 percent protein, bagged dairy goat feeds formulated for does in milk are a good choice for those who own only one or two milk goats. They take the guesswork out of formulating an adequate dairy ration.

Processing Milk for Kids

Fresh milk should be strained, even if you're feeding the milk to bottle kids. Milk filters can be purchased at most farm stores and they're better (and cheaper) than coffee filters from the supermarket. Once strained, milk can be fed as is to waiting kids, kept in the refrigerator up to a week, or processed for future use.

Freezing Milk

Freezing milk is the essence of simplicity. Simply pour it into gallon plastic milk jugs (if you'll be feeding it to multiple kids) or plastic soda bottles (to feed a single kid) and pop them in the freezer. Be sure to leave headroom for the milk to expand.

Canning Milk

Although home economists don't recommend canning milk, it can be done. Canned milk isn't tasty enough for our dinner table, but it works well for bottle-feeding kids. While I freeze milk rather than canning it, I've experimented with this recipe from Jerry Belanger's *Storey's Guide to Raising Dairy Goats* (North Adams, Mass.: Storey Publishing, 2001). The finished product separates and doesn't look nice on the shelf, but a good shake has it looking like milk again and bottle kids love it.

1. Follow your regular pressure-canner instructions regarding the amount of water to use, allowing steam to escape before closing the vent, and so on.
2. Fill the canning jars to within 1 inch of the top with fresh, warm (120°) milk.

3. Add sterilized lids and rings.
4. Process for 12 minutes at 12½ pounds pressure. Let the jars cool in the canner, undisturbed.
5. Remove jars and store in a dark place.

Another recipe calls for placing the jars in the canner in two or three inches of water, placing the cover on the canner, and turning the heat on high. After allowing steam to vent (without pressure) for ten minutes, put the pressure weight on the vent and bring the gauge up to five pounds of pressure, then turn off the heat and let the canner cool on its own. This milk is whiter than the other recipe, but I still prefer frozen milk.

Goat Milk for Human Consumption

The average dairy doe produces much more milk during a typical nine- to ten-month lactation than you'll probably want to put by for bottle kids. You could dry off your dairy doe, but why? Properly processed goat milk is wonderfully tasty and can be crafted into cheese, yogurt, ice cream, butter, and more. The key words are *properly processed,* which is not as simple as processing milk for bottle kids. For additional dairy goat information, refer to Resources.

Appendix F

So You Want to Show Meat Goats?

GOAT SHOWS ARE THE PERFECT VENUE to meet fellow goat breeders and publicize your stock. Opportunities range from county fair competitions to the big-time — a national Boer goat show.

If you plan to show at county fair level, quality meat goats of any breed will work just fine, but ask your county agent to make certain there are classes for that breed at your county fair.

Future Farmers of America and 4-Hers show goats of any breed, though most club projects are Boers or Boer crosses.

Showing Boer Goats

Most meat goat registries focus mainly on meat production, so if you want to show a lot and to work toward ennoblements (see page 28) or titles like Permanent Grand Champion, you must show Boer goats. Three Boer registries (American Boer Goat Association, International Boer Goat Association, and United States Boer Goat Association) sanction many shows each year and offer permanent award incentives for goats that win.

Each Boer registry's incentives program is slightly different, as are their showing rules and breed standards. To start, contact each organization to see which are active in the states where you plan to show. Join those clubs, request rule books, and study until you know protocols back and forth.

Compare their judging standards to your goats. Especially note traits each group heavily penalizes or considers elimination faults; if your goats have them, don't show those goats.

Study the Style

Fads come and go, so before selecting your show prospects, visit several shows. Note what sort of goats are winning. Do yours compare? As I write this, long-bodied, longer-legged, long-necked, unwrinkled wethers are winning in 4-H and FFA competition, so "wether sires" (bucks likely to sire that type of winning wether) are the darlings of most show judges.

These judges tend to give squat, heavily wrinkled, old-style, African-look bucks the gate. There are, however, judges who still favor original type, so if you have that type of goat, find out who those judges are, so you can exhibit at the shows where they judge.

Learning the Ropes

Strike up conversations with participants (goat people are notoriously friendly and most are happy to field beginners' questions), but do it when they aren't busy prepping or showing goats. And join a big, active email group like the_ boer_goat (see Resources) where you'll learn more about registry and showing "politics" than anyplace else on earth.

You must train your goats to walk smartly, stand still, and allow a strange judge to touch them. Meat goats are shown not in halters like cattle and horses, but with modified chain collars around their necks.

Learn to clip your goats to emphasize their stellar qualities and draw the judge's eye away from their faults. (See Appendix G.) Training and trimming takes practice, so ask a mentor to show you how and start preparing your goats well before their first show.

Keep Your Sense of Humor

Consider first shows a chance to learn things firsthand. If things go awry, so don't fret. Meat goats are powerful animals and sometimes do things their way, not yours. Consider the 300-pound ennobled buck who, weary of the show ring, simply exited the in-gate, hauling his very experienced and highly agitated handler alongside. Or the many who have deftly peed on the judge.

Things happen. Experienced show folks take them in stride. Showing is a breeders' showcase but it's also fun. You should enjoy it too.

Appendix G

Clipping for Shows

WHEN PREPARING YOUR GOATS FOR SHOWS, it's important to present them to their best advantage. Here are diagrams illustrating the most popular market wether (4-H) clip, as well as a detailed Boer show goat clipping diagram (from information provided by Claudia Marcus-Gurn of MAC Goats in Winona, Missouri).

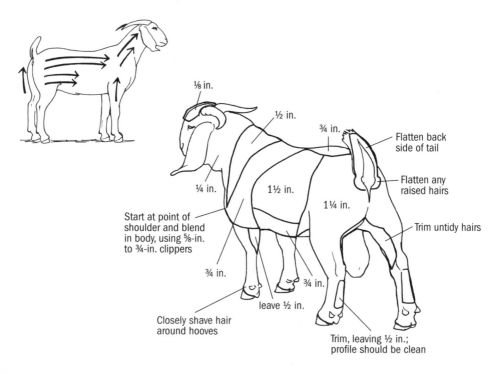

⅛ in.

½ in.

¾ in.

Flatten back side of tail

Flatten any raised hairs

¼ in.

1½ in.

1¼ in.

Trim untidy hairs

Start at point of shoulder and blend in body, using ⅝-in. to ¾-in. clippers

¾ in.

¾ in.

leave ½ in.

Closely shave hair around hooves

Trim, leaving ½ in.; profile should be clean

Appendix H

Emergency Euthanasia

EVEN IF YOU DON'T SLAUGHTER YOUR OWN GOATS or allow buyers to slaughter them on your farm, you need to know how to put one down. The most acceptable way is to shoot the animal. Here's how.

Use a .22-caliber long rifle, 9 mm, or .38-caliber gun. The muzzle of the gun should be held 4 to 12 inches away from the skull when fired. The use of hollow-point or soft-nose bullets will increase brain tissue destruction and reduce the chance of ricochet. When performed skillfully, euthanasia by gunshot induces immediate unconsciousness, is inexpensive, and does not require close contact with the animal. This method should be attempted only by individuals trained in the use of firearms and who understand the potential for ricochet. All humans and other animals should remain out of the line of fire.

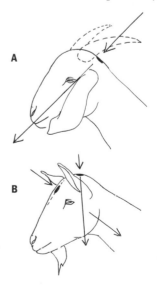

(A) Aim at the midline of the back of the goat's skull, just behind the bony ridge where the horns protrude, toward the back of the animal's chin.

(B) Kids less than four months of age or polled dairy goats may be shot from the front or top of the skull.

Resources

Meat Goat Producers

Featured in the Producers Profiles

Pat and Clark Cotton
Bending Tree Ranch
Damascus, Arkansas
501-679-4936
www.bendingtreeranch.com

**Beth and Randy Ellerbrock
and Spencer Toner**
Star E Ranch
Clayton, Illinois
217-894-6223
www.star-e-ranch.com

Mona and Joey Enderli
Enderli Farms
Baytown, Texas
281-421-8073

Al and Kirsten Kosinski
Black Bell Acres
Alton, Michigan
417-778-6009
www.blackbellacres.com

Dave and Judy Muska
Lazy J Goat Farm
Schulenburg, Texas

Janice and Roger Nodine
Southern Belle Farms
Portland, Tennessee
615-323-7138

Lee and Connie Reynolds
Autumn Farm Boers
Ravenswood, West Virginia
304-273-2610
www.autumnfarmboers.com

John and Sue Weaver
Dreamgoat Annie
Mammoth Spring, Arkansas
870-966-3569
www.dreamgoatannie.com

Organizations

Meat Goat Organizations

American Meat Goat Association
Sonora, Texas
325-387-6100
www.meatgoats.com

Canadian Meat Goat Association
Annaheim, Saskatchewan
306-598-4322
www.canadianmeatgoat.com

Boers

American Boer Goat Association
San Angelo, Texas
325-486-2242
www.abga.org

Boer Goat Breeders Association of Australia
Armidale, Australia
02-6773-5177
www.australianboergoat.com.au

British Boer Goat Society
info@britishboergoatsociety.co.uk
www.britishboergoatsociety.co.uk

International Boer Goat Association
Whitewright, Texas
903-364-5735
www.intlboergoat.org

New Zealand Boer Goat Breeders Association
boergoats@nzboer.co.nz
www.nzbgba.co.nz

United States Boer Goat Association
Spicewood, Texas
866-668-7242
www.usbga.org

Myotonics

International Fainting Goat Association
Darlington, Pennsylvania
724-843-2084
www.faintinggoat.com

Myotonic Goat Registry
Adger, Alabama
205-425-5954
http://myotonicgoatregistry.net

Kikos

American Kiko Goat Association
254-423-5914
www.kikogoats.com

International Kiko Goat Association
Bluff City, Tennessee
888-538-4279
www.theikga.org

Savanna

North American Savanna Association
Hallsville, Missouri
573-696-2550
http://savannahassociation.com

Spanish

Spanish Goat Association
The Plains, Virginia
540–687–8871
www.spanishgoats.org

Multiple Breeds and Types

Pedigree International
Humansville, Missouri
417-754-8135
www.pedigreeinternational.com
Breeders' directories; registers Kiko,
Gene-Master, Myotonic, Tennessee
Meat Goats, Savanna, Spanish, TexMaster, Sako, Cashmere, as well as pet and
pack type goats

Miscellaneous Goat Organizations

International Goat Association
Little Rock, Arkansas
501-454-1641
www.iga-goatworld.org

Sheep and Goat Organizations

Maryland Small Ruminant Page
www.sheepandgoat.com/organ.html
A comprehensive list of goat, sheep, and
related organizations

North American Packgoat Association
Boise, Idaho
www.napga.org

Sustainable Agriculture Organizations

American Grassfed Association
Denver, Colorado
877-774-7277
www.americangrassfed.org

**ATTR - National Sustainable Agricul-
ture Information Service**
Fayetteville, Arkansas
800-346-9140
www.attra.org

Center for Rural Affairs
Lyons, Nebraska
402-687-2100
www.cfra.org

Family & Small Farms Program
National Institute of Food and
Agriculture
Washington, D.C.
202-720-4423
www.nifa.usda.gov/familysmallfarms.cfm

National Organic Program
Agricultural Marketing Service
www.ams.usda.gov/AMSv1.0/NOP

**Sustainable Agriculture Research and
Education**
National Institute of Food and
Agriculture
Washington, D.C.
202-720-4423
www.sare.org

Veterinary Organizations

American Association of Small Ruminant Practitioners
Guthrie, Kentucky
270-483-2090
www.aasrp.org

American Holistic Veterinary Medical Association
Bel Air, Maryland
410-569-0795
www.ahvma.org

American Veterinary Medical Association
Schaumburg, Illinois
800-248-2862
www.avma.org

Directories

Breeders' World
Bellevue, Ohio
419-618-8267
www.breedersworld.com

Get Your Goats
http://getyourgoats.com

University-Generated Meat Goat Information and Publications

Cornell University Department of Animal Science
Ithaca, New York
607-255-2862
www.ansci.cornell.edu

Dairy and Animal Science Department
Penn State College of Agricultural Sciences
University Park, Pennsylvania
www.das.psu.edu

Department of Animal & Food Sciences
University of Kentucky
Lexington, Kentucky
859-257-2686
www.uky.edu/Ag/AnimalSciences

Mississippi State University Extension Service
Mississippi State, Mississippi
662-325-3034
http://msucares.com

Free (noncredit) Online Meat Goat Courses

Meat Goat Home Study Course
Penn State College of Agricultural Sciences
http://bedford.extension.psu.edu/agriculture/goat/Goat%20Lessons.htm

Web-based Training and Certification Program for Meat Goat Producers
Langston University
www.luresext.edu/goats/training/qa.html

Interactive Tools

BoerGoats.com OnLine Pedigree Generator
www.boergoats.com/tools/OnLinePedigree
This works for other breeds as well.

Medical Assessment Form
www.cometothefarm.com/medical_assessment.htm
Fill it out, print the form, and have all the information you need ready at hand when you call the vet.

Pedigree Generator
www.sitstay.com/pedi
Use this easy-to-use generator to write five-generation Web page pedigrees for any type of animal, including goats.

Recordkeeping Software

Agritec Software
www.agritecsoft.com
Ovitec

Goat Breeders Notebook
www.goatsoftware.com

HerdLogic
www.herdlogic.com

Ranch Manager: Goat Edition
Lion Edge Technologies
www.lionedge.com

Valuable Websites

Bar None Meat Goats
www.barnonemeatgoats.com
Scores of useful articles about meat goat husbandry and showing

Beyond the Sidewalk
http://beyondthesidewalk.com
The Web page of rural business marketing guru, Ellie Winslow. Subscribe to her free marketing newsletter; also click on Fun and Useful in the menu to access goat links and download farm marketing tips in PDF format.

BigNoseBird.Com
www.bignosebird.com
Everything you need to build great websites (for free)

The Biology of the Goat
www.goatbiology.com
Free goat gestation calculator software in PC and Mac versions, fantastic goat biology animations, and best-bet information on doing your own goat fecal exams

The Boer & Meat Goat Information Center
www.boergoats.com
Over 1,000 articles plus a comprehensive breeders directory, market results, production sale announcements, and more

Canada Plan Service
www.cps.gov.on.ca
Download hundreds of farm-related building plans for free.

Clear Creek Farms
www.motesclearcreekfarms.com
Easy-to-access articles covering every
conceivable subject relating to meat
goats

**E (Kika) de la Garza Institute for Goat
Research**
Langston University
www.luresext.edu/goats
Articles, fact sheets, interactive tools

Farming the Net
Ohio State University Extension
http://farmnet.osu.edu
All about the Internet, for farm-based
businesses

Fias Co Farm
http://fiascofarm.com
Much of the material at this truly outstanding 200-plus page dairy goat site
applies to meat goats too.

GoatWorld.com
www.goatworld.com
Scads of articles, USDA auction reports,
and home of both Goat 911 and the
Goat 911 Emergency Evacuation
Network

Jack and Anita Mauldin's Boer Goats
www.jackmauldin.com
Pages and pages of first-class goat
information

Maryland Small Ruminant Page
www.sheepandgoat.com
Thousands of up-to-date links on hundreds of sheep and goat topics

**New South Wales Department of
Primary Industries**
www.dpi.nsw.gov.au/agriculture
A wide range of useful articles about
goat raising

SheepGoatMarketing.info
http://sheepgoatmarketing.info
Directories, articles, news — a treasure
trove of goat marketing know-how

**Southern Consortium for Small
Ruminant Parasite Control**
www.scsrpc.org
Loads of information about parasite
control.

South African Boer Goats
www.sa-boergoats.com
A rich source of information no matter
what kind of goats you raise

Webmonkey
www.webmonkey.com
Website resources including many free
tutorials

Guardian Animals and Livestock Predation Information

Flock & Family Guardian Network
www.flockguard.org

Livestock Guardian Dogs
www.lgd.org

Predator and Wildlife Management
Maryland Small Ruminant Page
www.sheepandgoat.com/predator.html
Extensive links

Supplies

Goat-specific

Caprine Supply
De Soto, Kansas
913-585-1191
www.caprinesupply.com
Goat vaccines and equipment

Hoegger Goat Supply
Fayetteville, Georgia
800-221-4628
www.hoeggergoatsupply.com
Complete line of goat supplies (including dairy goat supplies and equipment), books, and more (for example, nylon goat harnesses, wagons, carts, and pack-goat gear)

Goat Handling Equipment

D-S Livestock Equipment
Frostburg, Maryland
800-949-9997
http://dslivestock.biz
Feeders, handling and containment equipment, and scales

Sydell, Inc.
Burbank, South Dakota
800-842-1369
www.sydell.com
Handling and containment equipiment

Tarter Farm and Ranch Equipment
Dunnville, Kentucky
800-733-4283
www.tartergate.com
Feeders and handling and containment equipment

Other

American Livestock and Pet Supply
Madison, Wisconsin
800-356-0700
www.americanlivestock.com
Pharmaceuticals, supplements, general farm and ranch supplies

Jeffers, Inc.
Dothan, Alabama
800-533-3377
www.jefferslivestock.com
Pharmaceuticals, supplements, general farm and ranch supplies

Molly's Herbals
Okemos, Michigan
molly@mollysherbals.com
http://fiascofarm.com/herbs
Herbal animal care products

Paper Direct
Colorado Springs, Colorado
800-272-7377
www.paperdirect.com
Home publishing supplies

Port-a-Hut Inc.
Storm Lake, Iowa
800-882-4884
www.port-a-hut.com
Complete line of Quonset-style animal housing

Premier1 Supplies
Washington, Iowa
800-282-6631
www.premier1supplies.com
Separate sheep and goat supplies, fencing, clippers, and shearing machines

Sullivan Supply Inc.
www.sullivansupply.com
Grooming tools and gadgets, blankets,
stands, shampoos, and conditioners for
show goats

Valley Vet Supply
Marysville, Kansas
800-419-9524
www.valleyvet.com
Pharmaceuticals, supplements, general
farm and ranch supplies

Goat Meat

American Meat Goat Association
Chevron Recipes
www.meatgoats.com/cookbook.pdf

Chevon Recipes
Copeland Family Farms, LLC
www.goatmeats.com

Cottamundra Kid
www.cootamundrakid.com.au
Recipes and goat meat information

Cooking Goat Information
Jack & Anita Mauldin's Boer Goats
www.jackmauldin.com
Goat meat recipes (lots of good ones!)

Periodicals

Goat Rancher
888-562-9529
www.goatrancher.com

Meat Goat Monthly News
Ranch Publishing
325-655-4434
www.ranchmagazine.com/mgn.html

Ranch & Rural Living
325-655-4434
www.ranchmagazine.com

United Caprine News
817-297-3411
www.unitedcaprinenews.com

Recommended Reading

Aubrey, Sarah Beth. *Starting & Running Your Own Small Farm Business*. North Adams, Massachusetts: Storey Publishing, 2007.

Belanger, Jerry. *Storey's Guide to Raising Dairy Goats*. North Adams, Massachusetts: Storey Publishing, 2010.

Damerow, Gail. *Fences for Pasture & Garden*. North Adams, Massachusetts: Storey Publishing, 1992.

———. *Your Goats: A Kid's Guide to Raising and Showing*. North Adams, Massachusetts: Storey Publishing, 1993.

Dawydiak, Orysia, and David E. Sims. *Livestock Protection Dogs: Selection, Care, and Training*. Loveland, Colorado: Alpine Publications, 2004.

Dohner, Jan Vorwald. *Livestock Guardians: Using Dogs Donkeys and Llamas to Protect Your Herd*. North Adams, Massachusetts: Storey Publishing, 2007.

Eddy, Carolyn. *Diet for Wethers*. Eagle Creek, Oregon. Eagle Creek Packgoats, 2001.

———. *Practical Goatpacking*. Eagle Creek, Oregon. Eagle Creek Packgoats, 1999.

Ekarius, Carol. *How to Build Animal Housing*. North Adams, Massachusetts: Storey Publishing, 2004.

———. *Small-Scale Livestock Farming*. North Adams, Massachusetts: Storey Publishing, 1999.

———. *Storey's Illustrated Breed Guide to Sheep, Goats, Cattle, and Pigs*. North Adams, Massachusetts: Storey Publishing, 2008.

Grandin, Temple. *Humane Livestock Handling*. North Adams, Massachusetts: Storey Publishing, 2008.

Grant, Lynella. *The Business Card Book: What Your Business Card Reveals about You — and How to Fix It*. Scottsdale, Arizona: Off the Page Press, 1998.

Green, J. S. and Woodruff, R. A. "Breed Comparisons and Characteristics of Use of Livestock Guarding Dogs." The Society for Range Management: Wheat Ridge, Colorado. *Journal of Range Management* 41, No. 3 (1988): 249-251. Available online at the University of Arizona Institutional Repository's Website: *www.uair.arizona.edu*

Haynes, N. Bruce. *Keeping Livestock Healthy.* North Adams, Massachusetts: Storey Publishing, 2001.

Klotz, Jennifer-Claire V. *How to Direct-Market Farm Products on the Internet.* Washington D.C.: United States Department of Agriculture, Agricultural Marketing Service, 2002. Out of print but you can download it free (it's a huge file) at PennState's Agricultural Marketing Website: *http://agmarketing.extension.psu.edu/Retail/howdrctmrktoninternet.html*

Knight, A.P. and R.G. Walter. *A Guide to Plant Poisoning in Animals in North America.* Jackson, Wyoming. Teton NewMedia, 2001.

Macher, Ron. *Making Your Small Farm Profitable.* North Adams, Massachusetts: Storey Publishing, 1999.

Mionczynski, John. *The Pack Goat.* Boulder, Colorado: Pruett Publishing, 1992.

Mitcham, Stephanie and Allison Mitcham. *Meat Goats: Their History, Management and Diseases.* Sumner, Iowa. Crane Creek Publications, 2000.

Popyk, Bob. *Here's My Card: How to Network Using Your Business Card to Actually Create More Business.* New York: St. Martin's Press, 2000.

Rigg, Robin. "Livestock Guarding Dogs: Their Current Use World Wide." Gland, Switzerland: IUCN/The Canid Specialist Group, 2001. Available online at The Canid Specialist Group's Website: *www.canids.org/occasionalpapers/livestockguardingdog.pdf*

Singh-Knights, Doolarie and Marlon Knights. *Feasibility of Goat Production in West Virginia: A Handbook for Beginners.* Morgantown, West Virginia: West Virginia University, 2005.

Smith, Mary C. and David M. Sherman. *Goat Medicine.* 2nd ed. New York: Wiley-Blackwell, 2009.

Solaiman, Sandra G. "Assessment of the Meat Goat Industry and Future Outlook for U.S. Small Farms." Tusegee, Alabama: Tuskegee University, 2007. Available online at Iowa State University's Agricultural Marketing Resource Center Website: *www.agmrc.org/commodities__products/livestock/ goats/meat_goats.cfm*

Tillman, Peggy. *Clicking with Your Dog: Step-by-Step in Pictures.* Waltham, Massachusetts; Sunshine Books, 2000.

Weathers, Shirley A. *Field Guide to Plants Poisonous to Livestock: Western U.S.* Fruitland, Utah; Rosebud Press, 1998.

Weaver, Sue. *Get Your Goat.* North Adams, Massachusetts: Storey Publishing, (forthcoming).

Winslow, Ellie. *The Complete Idiot's Guide to Raising Goats.* New York: Alpha Books, 2010.

———. *Economy Proofing Rural Business: Better Marketing Means More Money.* Reno, Nevada: Freefall Press, 2009.

———. *Making Money with Goats.* 5th ed. Reno, Nevada: Freefall Press, 2005.

———. *Marketing Farm Products: And How to Thrive Beyond the Sidewalk.* Reno, Nevada: Freefall Press, 2007.

———. *Growing Your Rural Business from the Inside Out.* Reno, Nevada: Freefall Press, 2008.

YahooGroups

Meat Goat Production

Commercial Goats1
http://finance.groups.yahoo.com/group/
CommercialGoats1

Goat and Sheep Rancher
http://finance.groups.yahoo.com/group/
GoatandSheepRancher

Goat Meat Production
http://pets.groups.yahoo.com/group/
GoatMeatProduction

Breeds

The Boer Goat
http://finance.groups.yahoo.com/group/
The_Boer_Goat

Boer Goats
http://tech.groups.yahoo.com/group/
boergoats

Colored Boer Goats
http://finance.groups.yahoo.com/group/
coloredboergoats

Kiko Meat Goat
http://tech.groups.yahoo.com/group/
Kikomeatgoat

Myotonic Goats
http://pets.groups.yahoo.com/group/
MyotonicGoats

Goat Emergencies/911

Goat ER
http://finance.groups.yahoo.com/group/
GoatER

Med-A-Goat 911
http://finance.groups.yahoo.com/group/
Med-A-Goat911

Working Goats

Cart/Wagon Goats
http://pets.groups.yahoo.com/group/
Cart_Wagon_Goats

Pack Goat
http://pets.groups.yahoo.com/group/
packgoat

Misc.

Goat Biology
http://tech.groups.yahoo.com/group/
GoatBiology

Goats 101
http://tech.groups.yahoo.com/group/
Goats_101

Goats for Sale
http://pets.groups.yahoo.com/group/
Goatsforsale

Home Dairy Goats
http://tech.groups.yahoo.com/group/
homedairygoats

Practical Goats
http://pets.groups.yahoo.com/group/
practical-goats

Livestock Guardians

Donkeys
http://tech.groups.yahoo.com/group/
Donkeys

Goats & Livestock Dogs
http://pets.groups.yahoo.com/group/
goatslivestockdogs

Llama Talk
http://pets.groups.yahoo.com/group/
llamatalk

Working LGDs
http://finance.groups.yahoo.com/group/
workingLGDs

Glossary

A

ABGA: American Boer Goat Association.

AKGA: American Kiko Goat Association.

ALBC: American Livestock Breeds Conservancy, a group dedicated to preserving and promoting rare and endangered breeds of American livestock and poultry.

abortion: Early termination of pregnancy.

abomasum: The third compartment of the ruminant stomach; the compartment where digestion takes place.

accelerated kidding: When a doe kids more often than once a year.

African Goat Flock Company: A New Zealand company that shipped frozen Boer goat embryos to Olds College in Canada during the early 1990s.

afterbirth: The placenta and any fetal membranes expelled from a doe after kidding; see also **placenta.**

American MeatMaker: A Kiko/Boer composite breed registered by the International Kiko Goat Association; American MeatMakers are predominately Kiko.

amino acid: One of the building blocks of protein.

ammonium chloride: A mineral salt fed to male goats to inhibit the formation of bladder and kidney stones.

anemia: Deficiency of red blood cells.

anestrus: The period of time when a doe is not having estrous (heat) cycles.

anthelmintic: A substance used to control or destroy internal parasites; a dewormer.

antibody: A circulating protein molecule that helps to neutralize disease organisms.

antitoxin: An antibody capable of neutralizing a specific disease organism.

APHIS: Animal and Plant Health Inspection Service of the U.S. Department of Agriculture.

artificial insemination (AI): A process by which semen is deposited within a doe's uterus by artificial means.

ATTRA: The National Sustainable Agriculture Information Service; managed by the National Center for Appropriate Technology (NCAT) and funded under a grant from the U.S. Department of Agriculture's Rural Business-Cooperative Service.

autogenous vaccine: Vaccine made from organisms collected from a specific disease outbreak (e.g., autogenous caseous lymphadenitis vaccine is manufactured using bacteria harvested from pus collected from the lanced abscess of an infected goat).

B

banding: Castration by the process of applying a fat rubber ring to a buckling's scrotum using a tool called an elastrator.

billy: (slang) An uncastrated male goat; the preferred term is *buck*; see also **buck.**

birth date: The actual date a kid was born.

birth weight: The weight of a kid taken within 24 hours of birth.

bleating: Goat vocalization; also referred to as calling; see also **calling.**

blind teat: A nonfunctional teat having no orifice.

bloat: Excessive accumulation of gas in a goat's rumen.

Body Condition Score: A value from 1 to 5 used to describe a goat's body condition.

Boer goat: A large, heavily muscled meat goat developed in South Africa and imported to North America during the 1990s.

BoKi: A half Kiko and half Boer composite breed registered by the International Kiko Goat Association.

bolus: A large, oval pill; also used to describe a chunk of cud.

booster vaccination: A second or set of multiple vaccinations given to increase a goat's resistance to a specific disease.

Bo-Se: An injectable prescription selenium supplement.

BOSS: Black sunflower seeds used as feed.

breech birth: A birth in which the rear end of the kid is presented first.

breed: Goats of a color, body shape, and other characteristics similar to those of their ancestors, capable of transmitting these characteristics to their own offspring.

broken mouth: A goat that has lost some of its permanent incisors, usually five or more years of age.

browse: Morsels of woody plants, including twigs, shoots, and leaves.

buck: An uncastrated male goat; see also **billy.**

buckling: An immature, uncastrated male goat; an uncastrated male kid.

buck rag: A cloth rubbed on the scent glands of a buck and presented to a doe to see if she is in heat.

butting: The act of a goat bashing another goat (or a human) with its horns or forehead.

C

cabrito: (Spanish) baby goat; meat from a milk-fed kid.

CAE: Caprine arthritis encephalitis.

calling: Goat vocalization; also referred to as bleating; see also **bleating.**

capretto: (Italian) Kid goat meat; a term commonly used in parts of Europe and in Australia.

caprine: Having to do with goats.

castrate: To remove a male's testes.

cattle panel: A very sturdy large-gauge welded wire fence panel; sold in various lengths and heights.

cc: Cubic centimeter; same as a milliliter (ml).

CD/T: Toxoid vaccine used to protect against enterotoxemia (caused by *Clostridium perfringens* types C and D) and tetanus.

cellulose: A component of plant cells that most animals, including goats, are unable to digest.

cervix: The section of a doe's uterus that protrudes into the vagina; it dilates during birth to allow kids to pass through.

chevon: Goat meat of any kind (similar in usage to *beef* and *pork*).

CL: Caseous lymphadenitis.

cloudburst: Pseudopregnancy; a false pregnancy in which cloudy fluid fills the uterus and no fetus is present.

CMPK: A calcium, magnesium, phosphorus, and potassium product used to correct calcium deficiency.

coccidiostat: A chemical substance mixed with feed, bottle-fed milk, or drinking water to control coccidiosis.

CODI/PCI: Camelids of Delaware/Pet Center International; CODI/PCI Boer goats descend in every line to goats imported by Jurgan Schultz, of Camelids of Delaware, in 1994.

colostrum: First milk a doe gives after birth; high in antibodies, it protects newborn kids against disease; sometimes incorrectly called *colostrums*.

condition: Amount of fat and muscle tissue on an animal's body.

concentrate: A high-energy, low-fiber, highly digestible feed such as grain.

corpus luteum: The progesterone-producing mass of cells that form once an ovum (egg) is released from the ovary.

cover: to breed (i.e., a buck covers a doe).

crossbreed: An animal resulting from the mating of two entirely different breeds.

cull: To eliminate from a herd or breeding program; or the eliminated animal.

curry goat: A tough, old market goat of any sex.

creep feeding: To provide supplementary feed to nursing kids.

cud: A ruminant's undigested food, which the animal regurgitates to be chewed and swallowed again.

D

dam: The female goat parent.

dental pad: An extension of the gums on the upper jaw; it is a substitute for top front teeth.

dehorning: The removal of existing horns.

Deccox: The brand name of Decoquinate.

Decoquinate: A coccidiostat sometimes added to feed to control coccidiosis.

deworm: The use of chemicals or herbs to rid an animal of internal parasites.

dewormer: An anthelmintic; a substance used to rid an animal of internal parasites.

disbud: To destroy the emerging horn buds of a kid via the application of a red-hot disbudding iron.

drench: To give liquid medicine by mouth; a liquid medicine given by mouth.

dressing percentage: The percentage of a live goat that remains as a carcass after slaughter.

dystocia: Difficult labor or delivery.

E

Easter kid: A plump, milk-fed, four- to six-week-old market kid; see also **hothouse kid.**

Eid: An annual Islamic festival; there are two major Eids in the Muslim calendar.

elastrator: A plierslike tool used to apply heavy, O-shaped rubber bands called elastrator bands to a kid's scrotum for castration.

emaciation: Loss of flesh resulting in extreme leanness.

embryo: An animal in the early stage of development before birth; a fertilized egg.

embryo transplant: Implantation of embryos into a surrogate mother.

energy: A nutrient category of feeds usually expressed as total digestible nutrients (TDN).

ennoblement: An award granted Boer goats by the American Boer Goat Association and the International Boer Goat Association; it's earned through inspections by approved inspectors and awards won at sanctioned shows and performance trials.

entire: An uncastrated male animal.

entero: A shortened, common name for enterotoxemia; see also **pulpy kidney.**

epididymis: Tubules that carry sperm from a buck's testicles to the spermatic cord.

estrogen: Female sex hormone produced by the ovaries; estrogen is the hormone responsible for the estrous cycle.

estrus: The period when a doe is receptive — that is, she will mate with a buck (she's "in heat") and can become pregnant.

estrous cycle: The doe's reproductive cycle.

ewe: A doe or female goat; used in countries where sheep terms are used to describe goats.

extra-label: The use of a drug for a purpose for which it isn't federally approved.

F

fainting goats: A common name for Myotonic goats; see also **Tennessee fainting goats** and **Texas Wooden Legs.**

fatten: To feed for increased weight gain prior to slaughter.

fecal egg count (FEC): The number of worm eggs in a gram of feces; sometimes written as eggs per gram (EPG).

feeder kid: A lean, weaned kid that's fed to gain weight in preparation for slaughter.

field shelter: A basic shelter with a roof and three sides.

finished kid: A plump but not fat market kid.

finishing: The act of feeding an animal to produce a desirable carcass for market.

fitter: Someone who fits and shows goats for another party.

fitting: Preparing a goat for show.

flehmen: A goat's curling of the upper lip in order to increase its ability to discern scent.

flushing: Increasing the amount of feed fed to a doe before and during the breeding season; the act of using drugs to super-ovulate a doe and harvest the resulting ova.

follicle stimulating hormone (FSH): A hormone produced in the pituitary gland and used to stimulate egg production by the ovary.

foot bath: A chemical mixture that goats walk through or stand in, used for the treatment of foot rot.

forage: Grass and the edible parts of browse plants that can be used to feed livestock.

forb: A broad-leafed herbaceous plant (e.g., curly dock, plantain, or dandelion).

free-choice: Available 24 hours a day, seven days a week; hay and mineral mixes are generally fed free-choice.

freshen: When a doe kids and begins to produce milk.

G

Genemaster: A composite Kiko/Boer breed registered by the American Kiko Goat Association.

genotype: The genetic makeup of an animal or plant.

gestation: The length of pregnancy.

Goatex Group LLC: The New Zealand consortium of large-scale breeders who developed the Kiko meat goat.

graft: A procedure in which a kid (or kids) is transferred to and raised by a dam that is not its own.

granny (or auntie) doe: A pregnant doe close to kidding who tries to claim another doe's newborn kid.

grain: Seeds of cereal crops such as oats, corn, barley, milo, and wheat.

gummer: An old goat that has lost all of its teeth.

H

Halal: Islamic dietary law that regulates the preparation of food.

hay: Grass mowed and cured for use as off-season forage.

heart girth: Circumference of the chest immediately behind the front legs.

heat: Estrus.

heterosis: The increased performance of hybrids over purebreds; see also **hybrid vigor.**

heritability: The degree to which a trait is inherited.

hermaphrodite: An animal with both male and female sex organs.

hothouse kid: A plump, milk-fed, four- to six-week-old market kid; see also **Easter kid.**

hybrid vigor: The increased performance of hybrids over purebreds; see also **heterosis.**

hypobiosis: The cessation of development of a gastrointestinal nematode.

hypothermia: A condition characterized by low body temperature.

I

IBGA: International Boer Goat Association.

IKGA: International Kiko Goat Association.

immunity: Resistance to a specific disease.

inbreeding: Mating closely related individuals such as father and daughter, mother and son, and full or half siblings, usually to fix type.

in kid: Pregnant.

in milk: Lactating.

International MeatMaker: A Kiko/Boer composite breed registered by the International Kiko Goat Association; American MeatMakers are predominately Boer.

intramammary infusion: Mastitis medicine inserted directly into a teat via its orifice.

intramuscular (IM): Into muscle; used in relation to how injections are given.

intravenous (IV): Into a vein; used in relation to how injections are given.

Islam: The religious faith of Muslim people.

J

jug: A pen approximately four feet by five feet where a doe and her kids are kept for the first 24 to 72 hours after kidding; see also **mothering pen.**

K

Katsikia: (Greek) Kid goat meat as sold in Greek markets.

ketones: Substances found in the blood of late-term pregnant goats suffering from pregnancy toxemia.

Kiko: A large, meaty goat developed by the Goatex Group in New Zealand and imported to the United States during the 1990s.

Kiko Pedigree: A performance-based American registry for Kiko goats.

kosher: Sanctioned by Jewish dietary laws which regulate the preparation of food.

L

lactated ringers solution: Sterile fluid injected into a dehydrated goat to rehydrate it.

lactation: The period when a doe is giving milk.

lamb: A kid or baby goat; used in countries where sheep terms are employed to describe goats.

Landcorp Corporation Ltd.: A New Zealand company that shipped frozen Boer goat embryos to Olds College in Canada in the early 1990s.

larva: Immature stage of an adult parasite; the term applies to insects, ticks, and worms.

legume: A plant such as alfalfa, clover, or lespedeza.

libido: Sex drive; the desire to copulate.

linebreeding: The mating of individuals sharing a common ancestor.

livestock broker: A goat dealer who sells goats on commission.

lutenizing hormone (LH): The hormone that triggers a doe's ovulation and stimulates the corpus luteum to secrete progesterone; it also stimulates testosterone production in bucks.

lymph: A clear, watery, sometimes faintly yellowish fluid derived from body tissues; it contains white blood cells and circulates throughout the lymphatic system.

lymph node: Any of the small bodies located along the lymphatic vessels — particularly on the neck behind the ears, farther down the neck, and in the flank area — that filters bacteria and foreign particles from lymph fluid.

M

Magic: A widely used, homemade energy supplement made by combining one part corn oil, one part molasses, and two parts Karo syrup.

mastitis: Inflammation of the udder.

meat packer: A slaughterhouse owner who contracts with producers to purchase goats.

milk letdown: Release of milk by the mammary glands.

ml: Milliliter; the same as a cubic centimeter (cc).

monensin: A coccidiostat sometimes added to feed to control coccidiosis; marketed under the brand name Rumensin, monensin is highly toxic to equines.

monkey mouth: Lower jaw is longer than the upper jaw and teeth extend forward past the dental pad on the upper jaw; see also **sow mouth** and **underbite.**

mothering pen: A pen approximately four feet by five feet where a doe and her kids are kept for the first 24 to 72 hours after kidding; see also **jug.**

motility: The ability of sperm to move themselves.

Muslim: A member of the Islamic faith.

Muslim kid: A moderately lean market kid; preferred by Muslim ethnic buyers.

mutton: The meat from an adult sheep; in some countries the meat of older goats is also called mutton.

myotonia congenita: The inherited neuromuscular condition that causes the major muscles of Myotonic goats to temporarily seize up.

Myotonic Goat Registry: An organization devoted to registering and promoting Myotonic goats.

Myotonics: The preferred name for fainting goats, Tennessee fainting goats, and Texas Wooden Legs; muscular meat goats carrying the gene for myotonia congenita.

N

nanny: (slang) A female goat; the preferred term is *doe*.

necropsy: A postmortem (after death) examination.

nematode: A type of internal parasite; a worm.

nymph: An immature insect and tick that lacks developed sex organs.

O

off feed: Not eating as much as usual.

omasum: The third part of the ruminant stomach; it's sandwiched between the reticulum and the abomasum.

oocyst: A minute pouch or saclike structure containing a fertilized cell of a parasite.

open doe: A doe that isn't pregnant.

orifice: The opening in the end of a functional teat.

orchitis: Inflammation of the testicle.

ovum: An egg (plural: ova) or oocyte.

overshot mouth: When the lower jaw is shorter than the upper jaw and the teeth hit in back of the dental pad; also called parrot mouth.

over the counter (OTC): Nonprescription drug.

ovulation: The release of an egg from the ovary.

oxytocin: A naturally occurring hormone important in milk letdown and muscle contraction during the birthing process.

P

paddock: A small, enclosed area used for grazing.

palpate: To examine medically by touch.

papers: A registration certificate.

parturition: The act of giving birth.

pathogen: An agent that causes disease, especially a living microorganism such as a bacterium or a virus.

pedigree: A certificate documenting an animal's line of descent.

Pedigree International: A business that maintains registration books for many different kinds of animals, including Tennessee Meat Goats, TexMasters, Savannas, Improved Spanish, Kiko, and Cashmere goats.

pelt: The skin of a goat with the hair left on.

percentage: A crossbred goat that is at least 50 percent a specific breed (e.g., Percentage Boer, Percentage Kiko).

perennial: A plant that doesn't die at the end of its first growing season but instead returns and regrows from year to year.

pH: A measure of the activity of hydrogen ions in a solution and, therefore, its acidity or alkalinity; refers to a solution's place on the pH scale.

pharmaceutical: A substance used in the treatment of disease; a drug, medication, or medicine.

phenotype: An individual's observable physical characteristics.

Philippino goat: A young female cull at a goat market.

pizzle: A stringy-looking structure at the end of a male goat's penis; see also **urethral process.**

placenta: Any fetal membranes expelled from a doe after kidding; see also **afterbirth.**

pneumonia: Infection in the lungs.

polled: A natural absence of horns.

postpartum: Refers to the time after giving birth.

prepartum: Refers to the time before giving birth.

predator: An animal that lives by killing and eating other animals.

probiotic: A living organism used to influence rumen health by assisting in the fermentation process.

progeny: Offspring.

progesterone: A hormone secreted by the ovaries and produced by the placenta during pregnancy.

prolific: Producing more than the usual number of offspring.

protein: A nutrient category of feed used for growth, milk, and repair of body tissue.

puberty: The time when a goat becomes sexually mature.

pulpy kidney: Another name for enterotoxemia.

purebred: An animal of a recognized breed kept pure for many generations — that is, not cross-bred with other breeds.

Q

quarantine: To isolate or separate an individual from others of its kind.

R

ram: An uncastrated male goat; word employed in countries where sheep terms are used to describe goats.

ration: Total feed given an animal during a 24-hour period.

registered animal: An animal having a registration certificate and number issued by a breed association.

rehydrate: To replace body fluids lost through dehydration.

recipient doe: A doe into which one or two flushed embryos are inserted and which she brings to term; also called a *recip.*

reticulum: The second chamber of a ruminant's stomach.

Roman-nosed: The convex profile of breeds such as the Boer meat goat and the Nubian dairy goat.

ringwomb: Failure of the cervix to open when a doe gives birth.

rotational grazing (or browsing): Moving grazing or browsing animals from one paddock to another before plant growth in the first is fully depleted.

roughage: Plant fiber.

rumen: The first compartment of the stomach of a ruminant, in which microbes break down the cellulose in plants.

Rumensin: The brand name for monensin.

ruminant: An animal with a multicompartment stomach.

rumination: The process whereby a cud or bolus of rumen contents is regurgitated, rechewed, and reswallowed; see also **cud.**

rut: The period during which a buck is interested in breeding females.

S

SARE: Sustainable Agriculture and Education; a program of the U.S. Department of Agriculture's Cooperative State Research, Education, and Extension Service (CSREES).

Savanna: A large, hardy meat goat from South Africa resembling a white Boer goat.

scouring: Having diarrhea.

scours: Diarrhea.

scrapie: The goat and sheep version of mad cow disease.

scrotum: The external pouch in which a buck's testicles are suspended.

scrub goat: A mixed-breed goat used for brush control.

seasonal breeder: A doe that comes into heat only during part of the year; most dairy goats are seasonal breeders.

seasonal market: The time of year when there is increased demand for a product.

selection: Choosing superior animals as parents for a future generation.

serology test: Blood test.

settle: Get pregnant.

sheath: The outer skin covering that protects a buck's penis.

shoebox kid: A newborn to ten-day-old market kid.

Shultz, Jurgan: An exotic animal importer who, with the aid of veteran South African Boer goat breeder Tollie Jordaan, imported the CODI/PCI Boer goats.

silent heat: In heat (estrus) but showing no outward signs of this condition.

sire: The male parent.

slaughter kid: A kid produced specifically for the meat market.

smooth mouth: A goat that has lost all of its permanent incisors, usually seven years of age or older.

sow mouth: The lower jaw is longer than the upper jaw and teeth extend forward past the dental pad on the upper jaw; see also **monkey mouth** and **underbite.**

Spanish goat: A type of meat goat common throughout the Southeast and Texas, said to descend from goats introduced to North America by Spanish explorers during the 1500s.

standing heat: The period during estrus (heat) when a doe allows a buck to breed her.

subcutaneous (SQ): Under the skin; used in relation to how injections are given.

sustainable agriculture: An approach to producing profitable farm products while enhancing natural productivity and minimizing adverse effects to the environment.

systemic: Affecting the entire body.

T

TDN: See **total digestible nutrients.**

tapeworm: A segmented, ribbonlike intestinal parasite.

teaser: A buck that has had its spermatic cords cut or tied (i.e., has had a vasectomy); these males cannot impregnate does but still have a sex drive.

Tennessee fainting goats: Another name for Myotonic goats; see also **Texas Wooden Leg.**

Tennessee Meat Goats: The trademark name for a large, meaty Myotonic goat breed developed by Suzanne W. Gasparotto, of Onion Creek Ranch in Lohn, Texas.

testosterone: A hormone that promotes the development and maintenance of male sexual characteristics.

Texas Wooden Legs: Another name for Myotonic goats; see also **fainting goats** and **Tennessee fainting goats.**

TexMaster: The trademark name for a composite breed developed by Suzanne W. Gasparotto; mostly Tennessee Meat Goat with a dash of Boer blood added to promote faster growth.

Tinsley, John: The itinerant farmworker who, in the 1880s, brought the first Myotonic goats to Tennessee.

Total Digestible Nutrients (TDN): A standard system for expressing the energy value of feed.

trace minerals: Minerals needed in only minute amounts.

trachea: Windpipe, leading from the throat to the lungs.

trimester: One-third of a pregnancy.

U

UC: Urinary calculi; mineral salt crystals ("stones") that form in the urinary tract and sometimes block the urethra of male goats.

udder: The female mammary system.

ultrasound: A procedure in which sound waves are bounced off tissues and organs in order to view internal structures; widely used to confirm pregnancy in does.

underbite: Lower jaw is longer than the upper and teeth extend forward past the dental pad on the upper jaw; see also **monkey mouth** and **sow mouth.**

urethral process: A stringy-looking structure at the end of a male goat's penis; see also **pizzle.**

USBGA: United States Boer Goat Association.

USDA: United States Department of Agriculture.

uterus: The female organ in which fetuses develop; the womb.

V

vagina: The passageway between the uterus to the outside of the body; the birth canal.

vascular: Pertaining to or provided with vessels; usually refers to veins and arteries.

vasectomied buck: A buck whose spermatic cords have been severed so that he cannot ejaculate sperm cells; see also **teaser.**

W

"Wearing his (or her) working clothes": A phrase used to suggest that an animal is in everyday pasture condition rather than having been spruced up or fitted for show.

wether: A castrated male goat.

withdrawal period: After receiving medication, the amount of time a producer must wait before sending an animal to market in order to ensure that no drug residues remain in its meat.

Y

yard: In British terminology, a drylot where animals (including goats) are kept.

yearling: A goat of either sex that is one to two years of age, or a goat that has cut its first set of incisors.

Z

zoonose: An animal disease that also infects humans.

Index

Page numbers in *italics* indicate illustrations or photos;
numbers in **bold** indicate charts or tables.

A

abomasum, *125, 126,* 127
abscess, 147, *147*
acidosis, 124, 129
advertising
  brochures and flyers, 256–58
  business cards, 253–56
  business names, 243, 247–48
  low-cost promotions strategies,
    258–60
  photography, 270–75, *272, 273, 275*
  publications, 253
  websites, 244–53
age categories, 61
Akbash dog, *187,* 187–88
Alpine dairy breed, 287, **288–89,** *290*
alternative medicine, 168
American Boki, 37, *38,* 41, 43
American Meatmaker, 37, 43
Anatolian Shepherd, 186–87, *187*
anatomy
  breed standards, 62–65
  buck's reproductive system,
    198–200, *199*
  digestive system, 124–29, *125, 126*
  doe's pelvic ligaments, 208, *208*
  doe's reproductive system, *204*
  esophagus (kids), *218*
  health indicators, 66–69
  horns, 67, 80
  lymph nodes, location of, *146*
  parts of a goat, 25, *26*
  teats and udders, 65, 67–69, *68*
  teeth, 66, *66*
  urinary tract problems and solu-
    tions, 202–4
Angoras, *40,* 40–41
antibiotics, 160–61, **160–61**
antitoxin, 159
auctions, 50–51, 227
Autumn Farm Boers, 22–24

B

barbed wire fencing, 117–18, *118*
barber-pole worm, 170–72
bedding, 104
behavior
  aggression, 79, *79*
  alarmed goats, 75–76, *76, 90,* 90–91

baby buckling, 201
breeding and kidding: does, 83–85
breeding: bucks, 81–82, *81–82*
changes prior to kidding, 209
feeding and range, 75–78, *76*
flehmen response, 81–82, *82*
newborn kids, 86–88, *88*
social order, 78–80, *79*
Bending Tree Ranch, 240
Black Bell Acres, 93–94
blankets for cold weather, *106,*
    106–7, *107*
blind zones, *90,* 90–91
bloat, 124, 144
bloodlines, 15–16
Boers, 10, 15–16
    breed fact sheet, 27, *27*
    breed standard conformation,
        56–61, *57*
    enobled, 28–29, *29, 56*
    overview, 25–31
    pedigree abbreviations and jar-
        gon, 70–71
    South African, *56,* 57–61, 71
BoKi, American, 37, *38,* 41, 43
bottle feeding, 220–222, *222*
bottle jaw, 144, *144*
breed standards
    back shape, 64, *64*
    dental, 62–63, *63*
    for Boers, 56–61
    forequarters, 63, *63*
    head shape and size, 62, *62*
    legs, 64, *64*
    testicles and udders, 65, *65*
breeders, evaluating, 45–51
breeding stock
    availability, 20
    Boers, *56,* 56–61
    buying, 45–51
    culling, 16, 57–60

fullblood bucks, 15
    marketing, 232
    meat goat traits, *62,* 62–65, *63,*
        64, 65
    prices for, 15, 18
    raising, 15–16
    selecting, 20, 54–72
    show goats, *56,* 56–61, 232–234
breeds. *See also* breeding stock;
    registries
    Angoras, *40,* 40–41
    availability of, *20*
    Boers, 10, 15–16, 25–31, *27, 29,*
        56–61, *57,* 71, 232–34
    climate considerations, 19
    composite, 37–38, *38*
    dairy goats, 39–40, *40,* 287–90,
        288–89
    Kikos, 10, 15, 32–34, *33*
    Myotonics, 10, 15, 34–36
    Savannas, 15, 31–32, *32*
    Spanish goats, 10, 38–39, *39*
    Tennessee Meat Goats, 15,
        36–38, *37*
brochures and flyers, 256–58
brown stomach worm, 172
brucellosis, 145
buck rag, 85
bucks
    anatomy, *26*
    baby buckling behavior, 201
    breeding behavior, *81,* 81–83, *82,*
        198–201
    commercial breeding stock, 15
    culling, 59
    head conformation, 62, *62*
    infertility, causes of, 199
    rams, compared, *87*
    reproductive system, 198–99, *199*
    safety when handling, 92–93,
        200–201, *201*

bucks *(continued)*

scent glands, 81, *81*
semen evaluation, 200
testicle standards, 65, *65*
business
cards, 253–56
names, 243, 247–48
buying goats, 45–51
evaluating sellers, 47, 49
locating sellers, 46
on the Internet, 48
production and dispersal sales,
49–50
sale barns, 50–51

C
cabrito, 1, 4
canning milk, 293
capretto, 4
caprine arthritis encephalitis
(CAE), 47, 55–56, 138, 145–46
caseous lymphadenitis (CL or
CLA), 47, 55–56, 138, *146,*
146–47
castration, 16, 268
catatonic state, 92
catch pen, 89, *89*
cattle or stock panels, 106–8, *108*
Caucasian Ovcharka dog, 188, *188*
chevon, 4
chutes, *89,* 89–90
clicker-training goats, 236
climate considerations, 19
clipping for shows, 298, *298*
coccidiosis, 147–48
cold protection
when transporting, 99
winter blankets, 106–7, *106–7*
colostrum, importance of, 216–17
composite breeds, 37–38, *38*
conversion table, 162
crossbred goats, 41–43

culling goats, 57–61

D
dairy goats, 39–40, *40,* 287–90,
288–89, *290*
Department of Environment, Food,
and Rural Affairs (DEFRA),
261–69
deworming, 6, 10, 47, 49
barber-pole worm, 170–72
brown stomach worm, 172
dewormer-resistant worms,
169–170
other gastrointestinal nematodes,
172, 173–74
reducing drug resistance, 173–74
digestive system, 124–29, *125, 126*
diseases. *See also* health care; parasites
abscess, 147, *147*
acidosis, 124, 129
bloat, 124, 144
bottle jaw, 144, *144*
brucellosis, 145
caprine arthritis encephalitis
(CAE), 47, 55–56, 138, 145–46
caseous lymphadenitis (CL or
CLA), 47, 55–56, 138, *146,*
146–47
coccidiosis, 147–48
enterotoxemia, 124, 148–49
floppy kid syndrome, 126, 149,
149
goat polio, 124, 150, *150*
hoof rot, 150–51
hypocalcemia, 151
Johne's disease, 47, 55–56, 138,
152
laminitis, 124
listeriosis, 124, 152–53, *153*
mastitis, 153

milk fever, 151
milk goiter, 145
pinkeye, 154
pizzle rot, 199
pneumonia, 55, 104, 154
pregnancy toxemia and ketosis,
    124, 154–55
scrapie, 156, *156*
soremouth, 156–57
tetanus, 157
urinary calculi, 124, 157–58
white muscle disease, 158
does. *See also* kidding
    anatomy, *26*
    breeding and kidding behavior,
        83–85
    caring for breeding does, 204–5
    cost, 18
    culling, 57–61
    nursing, 88, *88*, 126, *126*
    pelvic ligaments, 208, *208*
    reproductive system, *204*
    strutted udder, 84, *84*
    teat structure, 67–69, *68*
    udder and teats standards, 65, *65*
dog guardians, 185–94
donkey guardians, 183–84
drenching a goat, *166*, 166–68, *167*
    homemade vitamin drench, 167
dystocia, 210–14, *212, 213, 214*

E
ear tags, *280*, 280–81
electric wire fencing, 119–21, *120*
Enderli Farms, 72–74b
enobled Boers, 28–29, *29, 56*
enterotoxemia, 124, 148–49
epinephrine, 165–66
ethnic consumers, 1–3, 11–13, 228
    nonholiday meat preferences, **14**
euthanasia, 299, *299*

F
facilities. *See* housing
FAMACHA system, 175, *175*
feeding, 10–11. *See also* mineral
    supplements
    bottle feeding kids, 220–222, *222*
    breeding does, 205
    dairy doe, 293
    dietary suggestions, 123–24
    digestive system, 124–29, *125,*
        126
    feeders, *109*, 109–110, *110*
    forage, 124, 129, 131–35
    grain, 124, 129, 293
    hay, 131–35, 293
    pasture needs, 5, 10
    poisonous plants, 130–31
    tube feeding kids, 216–19, *217,*
        218, 219
    welfare code, 263–64
fencing, 6, 19, 113–22
    barbed wire, 117–18, *118*
    catch pen, 89, *89*
    cattle or stock panels, 106–8, *108*
    electric wire, 119–21, *120*
    handling facilities, 113
    head-in-fence syndrome, 116, *116*
    portable, *121*, 121–122, *122*
    posts, harvest your own, 117
    welfare code, 264
    woven wire, 114–17, *115*
first-aid kit
    farm-based, 140–41
    traveling, 98
flehmen response, 81–82, *82*
flight zones, *90*, 90–91
floppy kid syndrome, 126, 149, *149*
4-H projects, 17, 233
freezing milk, 293
fullblood goats, 41–42

G

Genemasters, 37, 43
goat 911, 139
goat milk
  for human consumption, 294
  for kids, 287–94
  processing, 293–94
goat polio, 124, 150, *150*
goat ranching, 10–11
goat-owner's checklist, 21
goat-producing states, **7**
goats. *See also* individual breeds
  anatomy, generally, 25, *26*
  magazines about, 45, 49
  physiology, 77
  pros and cons of raising, 5–7
  sheep, compared to, 86–87, *87*
  statistics, 3, **7,** 180
grain, 124, 129, 202, 293
grass-fed chevon, 231–32
Great Pyrenees, 189, *189*
guardians. *See* dogs by breed name;
    livestock guardians

H

halal, 1
handling, *89,* 89–93
  facilities, *89,* 89–90, *112,* 112–13
  flight zones and blind zones, *90,*
    90–91
  guardian dogs, *194*
  kids, *88*
  leading *vs.* driving, 91, *91*
  safety, 92–93, 200–201, *201*
harness goats, *238,* 238–39, *239*
hauling goats, 99–102
  dog crates, 99–100
  goat totes, *100,* 100–101
hay, 131–35, 293
  price per ton, 133
  quality, 132, 134

selecting, 131–32
  storage, 134
head-in-fence syndrome, 116, *116*
health care. *See also* diseases; kid-
    ding; mineral supplements,
    parasites
  alternative medicine, 168
  antibiotics, 160–61, **160–61**
  clinical signs of nematodes, **172**
  deworming, 6, 10, 47, 49, 169–
    174
  digestion, 124–29, *125, 126*
  drenching, *166,* 166–68, *167*
  drug resistance, 169–70, 173
  epinephrine, 165–66
  first-aid kit, 140–41
  healthy *vs.* sick behavior, **55**
  homemade vitamin drench, 167
  hooves, how to trim, *282,* 282–
    86, *283, 285*
  quarantine, 138
  records and documentation, 49
  shots, 9, 162–64, *163, 164*
  strategies, 136–38
  vaccinations, 10, 47, 48, 138,
    158–59
  veterinarians, 7–10
  vital signs, 140–43, *142, 143*
  welfare code, 263
herding, 79–80, 91, *91*
hoof rot, 150–51
hooves, how to trim, *282,* 282–86,
    283, 285
horns
  facts about, 80
  head-in-fence syndrome, 116, *116*
  health indicators, 67
  safety when handling, 92–93, *93*
housing and facilities, 5, 103–4
  feeders, *109,* 109–110, *110*
  fencing, 19, 106–8, *108,* 113–22

mini-shelters for kids, 105, *105*
pasture, 5, 10, 19
pens, 105–6
shelter, 19, 103–4, *104*
welfare code, 265–66
hypocalcemia (milk fever), 151

I

identification, 276–81
 ear tags, *280*, 280–81
 tattoos, 276–79, *278*, *279*
 temporary on-farm ID, 281, *281*
immunoglobulin replacers, 217
infertility, 199
International Meatmaker, 37, 43
Internet, 46, 48, 275. *See also* website design
intestines, small and large, *125*, 128

J

Johne's disease, 47, 55–56, 138, 152

K

Kangal dog, 189–90, *190*
Karakachan, *190*, 190–91
kidding, 5
 behavior of does, 85
 delivery, difficult or abnomal
  (dystocia), 210–14, *212*. *213*, *214*
 delivery, normal, 210, *210*
 kidding kit, 206–7
 preparing for, 206–7
 signs of impending kidding, 84,
  84, 207–9
 welfare code, 266
kids
 after birth, 214–15
 baby buckling behavior, 201
 bottle feeding, 220–222, *222*
 coats and sweaters, 107, *107*
 colostrum, importance of, 216–17

digestive system, 126, *126*
 mini-shelters, 105, *105*
 newborn behavior, 86–88, *88*,
  126, *126*
 nursing, 88, *88*, 126, *126*
 tube feeding, 216–19, *217–19*
Kikos, 10, 15, 32–34
 breed fact sheet, 33, *33*
Kinders, 287
Komondor (Hungarian sheepdog),
 191, 191–92
Kuvaszok, 192, *192*

L

LaBoer dairy doe, 40, *40*
labor needs, 19
LaMancha dairy breed, 40, 287,
 288–89, *290*
laminitis, 124
Lazy J Goat Farm, 195–97
lice, 178–79
listeriosis, 124, 152–53, *153*
liver fluke, 176
livestock guardians, 180–95
 dogs, 10, 185–94
 donkeys, 183–84
 llamas, 184–85
 predator behavior, 181–82
livestock transporter, 101–2

M

MAC goats, *6*, *29*, 43, 245
Maremma sheepdog, 193, *193*
market goats, 11–15, 226–32
 terminology, 4, 227
market kids, 227
marketing. *See also* advertising
 breeding stock, 232
 grass-fed and natural chevon,
  231–32
 market goats, 11–15, 226–32

marketing *(continued)*

  organic goat meat, 228–31
  show goats, 232–34
  working goats, *234,* 234–40, *235,*
    238, 239, 240
mastitis, 153
meat
  calorie, fat, and protein content,
    228
  packers, 227
  prices, 8
  sales, 228
  terminology, 4
meat goat traits, *62,* 62–65, *63, 64, 65*
Meatmakers, 37, 43
meningeal worm, 176
metric conversion table, 162
milk fever, 151
milk for kids, 287–94
  dairy breeds, 287–91, **288–89,** *290*
  processing milk, 293–94
milk goats. *See* dairy goats
milk goiter, 145
milking a goat, 291–92, *292*
milking codes, 267
mineral supplements
  disease prevention, 148, 151, 158,
    202
  feeding, 124, 135
  for breeding does, 205
  for dairy does, 293
  homemade drench, 167
mites, 179
Myotonia congenita, 35
Myotonics, 10, 15, 34–36
  breed fact sheet, 36, *36*

N

natural chevon, 228–29, 231–32
nematodes, 170–172, **172,** 173–74
Nigerian Dwarf goats, 287
nose bots, 178

Nubian dairy breed, 39–40, *40,*
    287, **288–89,** *290*

O

Oberhaslis dairy breed, 287,
    288–89, *290*
omasum, *125, 126,* 127
organic chevon, 160, 228–32
  certification standards, 230–31
Ozark Goat Trek, 43–44

P

pack goats, *234,* 234–37, *235*
parasites, 169–79
  barber-pole worm, 170–72
  brown stomach worm, 172
  dewormer-resistant worms, 170
  FAMACHA system, 175, *175*
  lice, 178–79
  liver fluke, 176
  meningeal worm, 176
  mites, 179
  nose bots, 178
  other gastrointestinal nematodes,
    172, 173–74
  tapeworm, 176
pasture, 5, 10
pedigree
  abbreviations, 70
  jargon, 71
pelvic ligaments, 208, *208*
pens, 105–8, *108*
  handling, 89, *89*
  mini-shelters for kids, 105, *105*
percentage goats, 41–42
pH
  of a goat's rumen, 129
  scale examples, 127
photographer, traveling, 240, *240*
photographing your goats, 270–75,
    272, 273, 275

pinkeye, 154
pizzle rot, 199
pneumonia, 55, 104, 154
poisonous plants, 130–31
Polish Tatra sheepdog, *194*, 194–95
portable fencing, *121*, 121–122, *122*
predators, 6, 19, 180–82
pregnancy toxemia and ketosis, 124, 154–55
Pritchard teat, 222, *222*
probiotics, 161
processing goat milk, 293–94
producer profiles
- Autumn Farm Boers, 22–24
- Bending Tree Ranch, 240
- Black Bell Acres, 93–94
- Enderli Farms, 72–74
- Lazy J Goat Farm, 195–97
- Ozark Goat Trek, 43–44
- Southern Belle Farms, 51–53
- Star E Ranch, 224–25

producers, evaluating, 45–51
production and dispersal sales, 49–50
promotion. *See* advertising
purebred goats, 41–42
PVC mineral feeders, *110*, 110–11
Pygmy goats, 287
Pyrenean Mountain dog, 189, *189*

Q

quarantine, 138

R

rabies, 152–53
range behavior, 76–79
registered goats, 41–42
registration papers, 49, 69–71
registries
- Boers, 30
- composite breeds, 38
- generally, 41–42
- Kikos, 33
- standards set by, 56–61
- Tennessee Meat Goats, 37

religious holiday sales, 1–2, 11–13, **12–13**
reticulum, 125, *125*
rumen, 125, *125*

S

Saanen dairy breed, 287, **288–89,** *290*
Sable dairy breed, 287, **288–89**
safety when handling bucks, 92–93, 200–201, *201*
sale barns, 50–51
Savannas, 15, 31–32, *32*
scrapie, 156, *156*
selenium, 158
semen evaluation, 200
sheep similarities and differences, 86–87, *87*
shoebox kids, 227
shots, 9, 162–64
- epinephrine, 165–66
- injection sites, *164*
- preparing the syringe, *163*

show stock
- clipping for shows, 298, *298*
- culling, 16, 57–60
- enobled Boer goats, 28–29, *29*
- marketing, 232–34
- raising, 16
- showing Boer goats, 295–97, *296*

slaughtering, 227–28
social order, 78–80
soremouth, 156–57
South African Boer Goat, *56*, 71, 232–34
- breed standard conformation, 57–61

Southern Belle Farms, 51–53

Spanish goats, 10, 38–39, *39*
Star E Ranch, 224–25
suckling kids, 227

T
tapeworm, 176
tattoos, 276–79, *278*, *279*
teat structure, 67–69, *68*
teeth, 66, *66*
Tennessee Fainting Goats. *See*
    Myotonics
Tennessee Meat Goats, 15, 36–38
    breed fact sheet, 37, *37*
tetanus, 157
tethering, 264–65
TexMasters, 37, *38*, 43
thiamine, 150
Toggenburg dairy breed, 287,
    288–89, *290*
toxoid, 159
transporting goats, 95–102
    cold protection, 99
    first-aid kit, 98
    hauling conveyance, 99–102
    heat exhaustion, 97–98
    planning ahead, 96–97
    stress factors, 95
    working with a livestock trans-
        porter, 101–2
tube feeding, 216–19, *217*, *218*, *219*

U
udder
    fill, 207
    standards, 65, *65*
    strutted, 84, *84*, 208
urinary calculi, 124, 157–58

V
V-shaped feeders, 109, *109*

vaccinations, 10, 47, 48, 138,
    158–59
    at birth, 217
    for breeding does, 205
    toxoid *vs.* antitoxin, 159
veterinarians
    being your own vet, 138–39
    finding and working with, 8–10,
        136
vision, 78
    flight zones and blind zones, *90*,
        90–91
vital signs
    heart rate and respiration, 143,
        143
    taking a goat's temperature,
        141–143, *142*

W
water, drinking, 77, 263
watering devices, *111*, 111–12
website design, 244–53
weighing your goat, **177**
wethers, 17
white muscle disease, 158
wild bezoar goat, 2
winter blankets, *106*, 106–7, *107*
working goats, *234*, 234–40, *235*,
    238, *239*, *240*
woven wire fencing, 114–17, *115*

Y
youth goat projects, 17, 233

STOREY'S GUIDE TO RAISING SERIES

For decades, animal lovers around the world have been turning to Storey's classic guides for the best instruction on everything from hatching chickens, tending sheep, and caring for horses to starting and maintaining a full-fledged livestock business. Now we're pleased to offer revised editions of the Storey's Guide to Raising series — plus one much-requested new book.

Whether you have been raising animals for a few months or a few decades, each book in the series offers clear, in-depth information on new breeds, latest production methods, and updated health care advice. Each book has been completely updated for the twenty-first century and contains all the information you will need to raise healthy, content, productive animals.

Storey's Guide to Raising BEEF CATTLE (3rd edition)

Storey's Guide to Raising RABBITS (4th edition)

Storey's Guide to Raising SHEEP (4th edition)

Storey's Guide to Raising HORSES (2nd edition)

Storey's Guide to Training HORSES (2nd edition)

Storey's Guide to Raising PIGS (3rd edition)

Storey's Guide to Raising CHICKENS (3rd edition)

Storey's Guide to Raising DAIRY GOATS (4th edition)

Storey's Guide to Raising MEAT GOATS (2nd edition)

Storey's Guide to Raising DUCKS (2nd edition)

Storey's Guide to Raising MINIATURE LIVESTOCK (NEW!)

Storey's Guide to Keeping HONEY BEES (NEW!)

Storey's Guide to Raising TURKEYS

Storey's Guide to Raising POULTRY

Storey's Guide to Raising LLAMAS

Other Storey Titles You Will Enjoy

Basic Butchering of Livestock & Game, by John J. Mettler Jr., DVM.
Clear, concise information for people who wish to slaughter their own meat for
beef, veal, pork, lamb, poultry, rabbit, and venison.
208 pages. Paper. ISBN 978-0-88266-391-3.

How to Build Animal Housing, by Carol Ekarius.
An all-inclusive guide to building shelters that meet animals' individual needs:
barns, windbreaks, and shade structures, plus watering systems, feeders, chutes,
stanchions, and more.
272 pages. Paper. ISBN 978-1-58017-527-2.

Humane Livestock Handling, by Temple Grandin with Mark Deesing.
Low-stress methods and complete construction plans for facilities that allow small
farmers to process meat efficiently and ethically.
240 pages. Paper. ISBN 978-1-60342-028-0.

Livestock Guardians, by Janet Vorwald Dohner.
Essential information on using dogs, donkeys, and llamas as a highly effective,
low-cost, and nonlethal method to protect livestock and their owners.
240 pages. Paper. ISBN 978-1-58017-695-8.
Hardcover. ISBN 978-1-58017-696-5.

Small-Scale Livestock Farming, by Carol Ekarius.
A natural, organic approach to livestock management to produce healthier
animals, reduce feed and health care costs, and maximize profit.
224 pages. Paper. ISBN 978-1-58017-162-5.

Starting & Running Your Own Small Farm Business, by Sarah Beth Aubrey.
A business-savvy reference that covers everything from writing a business plan
and applying for loans to marketing your farm-fresh goods.
176 pages. Paper. ISBN 978-1-58017-697-2.

Storey's Illustrated Breed Guide to Sheep, Goats, Cattle, and Pigs, by Carol Ekarius.
A comprehensive, colorful, and captivating in-depth guide to North America's
common and heritage breeds.
320 pages. Paper. ISBN 978-1-60342-036-5.
Hardcover with jacket. ISBN 978-1-60342-037-2.

Successful Small-Scale Farming, by Karl Schwenke.
An inspiring handbook to introduce the small-farm owner to the real potential
and difficult realities of living off the land.
144 pages. Paper. ISBN 978-0-88266-642-6.

These and other books from Storey Publishing are available
wherever quality books are sold or by calling 1-800-441-5700.
Visit us at *www.storey.com.*